WITHDRAWN

Paints, Inks, and Dyes

PAINTS, INKS and DYES

by Richard B. Lyttle

HOLIDAY HOUSE • NEW YORK

Printed in the United States of America

Library of Congress Cataloging in Publication Data

Lyttle, Richard B.
Paints, inks, and dyes.

SUMMARY: A history of paints, inks, and dyes—their origins in prehistoric times, and their uses then and now.
1. Colors—Juvenile literature. [1. Colors]
I. Title.
ND1510.L97 667 73-16876
ISBN 0-8234-0240-1

For INEZ STORER, an artist
whose enthusiasm for teaching
inspired this book

Acknowledgments

Books collect friends. This book began collecting many friends long before work began. Warm thanks for their time, encouragement, and sound advice go to Mr. and Mrs. Donald Burleson, Mr. and Mrs. C. Richard Roth, Mr. and Mrs. Lawrence Walters, Christine Nielsen, Michael Conway, Dr. Keene O. Haldeman, Edward F. Dolan Jr., Barthold Fles, Rickey Adams, my daughter, Jennyfur, and of course, my wife, Jean, who takes more than a friendly interest in all my books.

Special thanks go to Inez Storer, to whom this book is dedicated. Without her assurance of help, this book would not have been attempted.

Organizations assisting with information and pictures were National Paint and Coatings Association, Graphic Arts Technical Foundation, National Paint, Varnish and Lacquer Association, Paint Industry Education Bureau, A. B. Dick Company, Allied Chemical Corporation, Publishers' Development Corporation, *Screen Printing*, MacNair-Dorland Company, Sherwin-Williams Company, Glidden-Durkee Division of SCM Corporation, and Smithsonian Institution Press. Unless other credit is given, the author is responsible for the photographs.

Contents

1

The Bright Life

All too often we take our surroundings for granted. Because they have always been there, we fail to appreciate life's most valuable gifts. This is particularly true of color.

Can you imagine a world without color? Suppose that after seeing nothing but dull, gray tones for years, our eyes suddenly began seeing blue sky, green grass, golden flowers, red and purple sunsets and many-hued, vibrant rainbows. Then we would know color for what it is—a miracle.

It does seem a miracle that our eyes can see color at all. It is also a mystery. Scientists cannot yet explain exactly how it happens. In fact, they are not even certain about their understanding of light itself. Yet despite the mystery of sight, despite our imperfect knowledge, we have mastered color. We make it work and play at our command.

The achievement did not come easily. It took centuries of trial and error, and many of our greatest advances have been very recent.

A century ago, only people in comfortable circumstances could afford to paint their homes. Little more than a century ago, all dyes had to be laboriously extracted from plant or animal tissues. And it was hardly more than five centuries ago when a man combined ink and ingenuity to give the Western world its first book printed from movable type. In 1943 the ballpoint pen was a novelty, and a rather unsuccessful one at that.

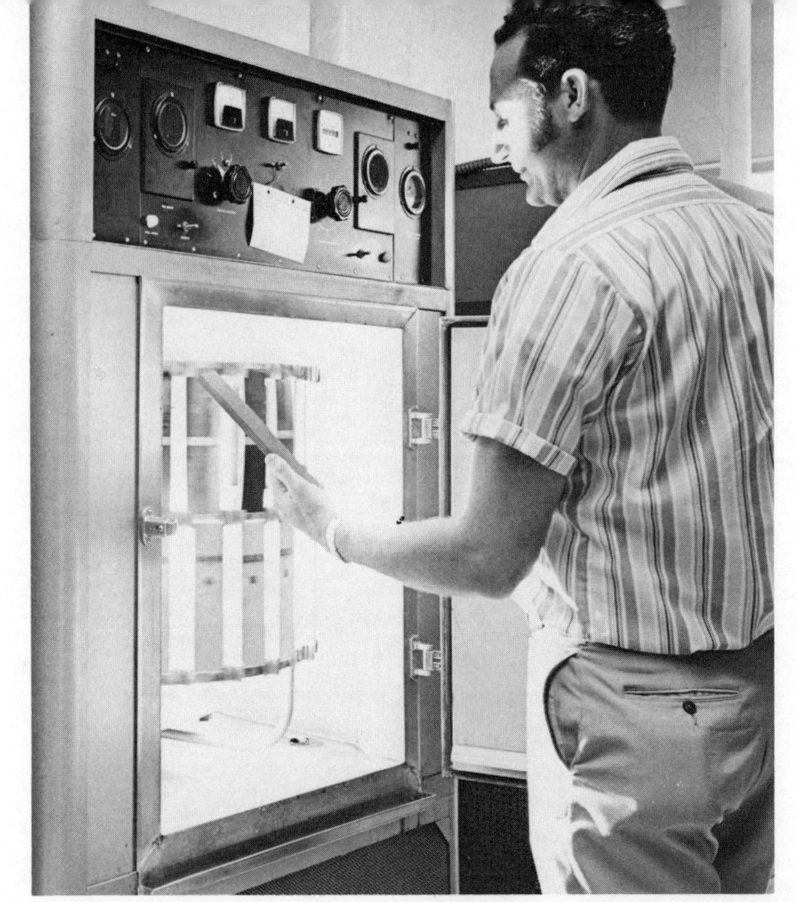

GLIDDEN-DURKEE DIV. OF SCM CORP.

Test panels in this cabinet show how well various paints hold up against salt spray. Modern paints are made and tested in a scientific manner today, a far cry from the hit-or-miss methods of centuries ago.

Today, thanks in large part to an expanded knowledge of chemistry, we can match any color in the rainbow. We can make paints that change from liquid to armor-tough coatings within a few minutes of application. We can make inks that mass-produce copies of paintings or colored photos with such fidelity it is difficult to distinguish them from their originals. We can make fadeproof and washproof dyes of any color, dyes that are bound so closely to the cloth that we cannot destroy their color without destroying the cloth itself.

Still, most of us take these marvels for granted.

Without paint to shield against rust, construction of bridges, skyscrapers, and steel ships would be impractical. In some climates, wooden homes would not last a lifetime if they were left unpainted. Without fast-drying inks, mass communication through high-speed printing would be impossible. Without writing inks of any kind, we would be little removed from the primitive state. Without dyes, we would be cloaked in monotony, both figuratively and actually. Self-expression through choice of clothing and furnishings would be blunted, to say the least.

Paints, inks, and dyes are not just important; they are a vital part of life. Looking back, we find them among the foundation stones of civilizations. Obviously, mankind's interest in paints, inks, and dyes began a long, long time ago.

A *Parisian ink vendor in the 1800s, with ink bottled for the customer while he waits.*

COLOR TRADITIONS

In prehistoric times, colors played a far different role than they do today, but they were no less important. Colors were symbolic, part of the spiritual or mystic life. They guarded against evil. They brought luck. They cured the sick. Often they served to explain the unknown.

Navajo Indians of the American Southwest, for instance, once believed that black mountains rose from the north to cause night. Blue mountains rose from the south to bring day. White mountains rose from the east to bring dawn, and yellow mountains rose from the west to bring twilight. Various other peoples, even as far away as China and Tibet, connected colors with directions. Sometimes good and evil were involved. The red god of the west was evil while the white god of the east was good. For many tribes, black meant death. Red stood for life, success, or good fortune. Others looked on green as the symbol of life or rebirth. When Cherokee Indians wanted rain, they carried green sticks in a ceremonial dance.

Among stone-age tribes, the use of red in burials was almost universal. Red spread on the body hid the pallor of death. Red probably stood for blood, the vital fluid of life. Archeologists tell us they think that the use of red in graves strongly indicates a belief by these primitive peoples in some kind of life after death. Some of these graves are 100,000 years old. The trace of red on prehistoric bones thus tells us that religion is a very ancient tradition.

Red served other purposes. After killing his quarry, the primitive hunter smeared his body with red to guard against the animal's avenging spirit. Sometimes the blood of the victim served as body paint, but often the red came from other sources. Lapland hunters, returning from a successful bear hunt, received a strange welcome. Their women and children spat upon them with a red juice they had prepared by chewing twigs of a certain plant. This prevented the spirit of the bear from harming the hunters. In other regions, where it was available, red clay served the same purpose.

Body paint had other uses. The African warrior, after killing an enemy, painted one side of his own body white and the other side red. This served to advertise his success in battle. Before a hunt or a battle, primitives painted their bodies to bring luck or to frighten their foes. The paint may have had an incidental effect in camouflaging the body, which contributed indeed to "good luck."

Primitives also used body paint in the simple belief that it improved their looks. This belief prevails among modern "tribes" and keeps the cosmetics industry alive and healthy to this day.

Archeologists are certain that paints and dyes were first used to color the body; but steadily, almost from the beginning, they spread to other things. A paint-covered palm pressed by accident or intent against a tree trunk or cave wall left a hand print, a distinctive mark or "signature." From this it was no great leap to drawing symbolic designs with a fingertip or stick wet with paint. The colors used in such designs were important.

As the first great civilizations rose in fertile regions of Africa, Asia, and the Middle East, colors gained more importance. The Egyptian warrior had to wear something purple to give him courage. If he could afford it, he wore an amethyst hanging from his neck. For the Chinese, gods were white, goblins red, and devils black. In the Middle East and many other regions, blue guarded against the "evil eye," that vile unknown something that caused sickness and injury, ill fortune, shame, and death. The leading camel in every desert caravan was always well decorated with blue.

For the Greeks, who tried to sort the world out in logical

divisions, blue stood for both earth and mankind. Red stood for fire and the human spirit. Green symbolized water. Yellow represented air. White clothed the gods.

The power of color guided the development of medicine. At first, herbs and salves were chosen mainly for their color. It was a happy accident that some of these preparations actually did have beneficial properties. Because jaundice turned the body yellow, yellow medicine served as the cure. Red bandages were needed to stop bleeding. In medieval Europe, a red sickroom was considered the best cure for smallpox. Belief in the curing power of color has continued for centuries. Less than 100 years ago, a large cult in the United States existed on the belief that various colored lights could cure the sick and assure good health. We cannot be too critical of this notion. Psychiatrists say that some colors do indeed soothe the troubled mind. Today's hospital rooms are often painted in pale green or other soft tones. Such surroundings actually can help to speed a patient's recovery.

Color, as we have already seen, was important in religion. The white of the Greek gods originated with sun worship. Yellow, gold, and white marked the deities in nearly all early religions. Ra ruled Egypt in yellow splendor. Hindus envisioned Vishnu in yellow raiment. Druids worshiped the golden bough found in the forests of Britain, actually nothing more than a branch of mistletoe, which still serves in our Christmas tradition as an excuse for kissing. There were other important colors in religion. For Mohammedans, green was most important because it stood for mother earth. Early Hebrew literature assigned red for sacrifice, sin, and love. Blue stood for glory. Purple meant splendor. White was the color of purity and joy. For Christians, yellow stood for Christ, red for the concept of the Holy Ghost, and blue for God.

Color symbolism influenced religious art and architecture, and much of the symbolism survives to this day in religious ritual. In many Christian churches red marks the Christmas season. Whites and yellows appear at Easter. Purple is used during Lent. Black is the color for funeral services.

Heraldry, a complex system of design and color which is still followed in England, rose with the powerful families of medieval times. In heraldry, gold and yellow stand for loyalty, silver for purity, blue for piety and sincerity, black for grief and penitence, green for hope and youth, purple for royalty, and orange for endurance.

In our country, heraldry was abandoned with the Revolution, but in academic life another old tradition remains. When today's college graduates receive their degrees, the costumes for the ceremonies are black robes trimmed with specific colors: red for theology, blue for philosophy, purple for law, green for medicine, yellow for science, and orange for engineering. Quaint? Maybe it is. But anyone who has worked hard to win a blue ribbon in a county fair bake sale or art show, or trained for months to vie for the blue ribbon in a track meet is in no position to make fun of traditions that use colors as symbols.

TRIAL AND ERROR

Through the centuries, while color has been important for one reason or another, it has not always been easy to find colors that worked. Man searched constantly for new and better colors. There were many failures. Sometimes the search led to odd dealings.

Artists of Europe at one time favored a delicate brown called "mummy," but popularity of the color dropped abruptly when it was discovered that the brown actually was mummy. Robbers had been taking bodies from the tombs of Egypt and grinding them up to produce paint. India once had a thriving trade in "Indian yellow." This paint, too, lost its appeal when it was learned that it came from the urine of cows that fed on mango leaves.

Though the sources of color were usually not as bizarre

as this, we shall see in subsequent chapters that craftsmen tried almost anything to acquire more colors. Through trial and error, guidelines gradually developed. There were rules governing color. Early in their experimentation, craftsmen learned to distinguish between pigments and dyes.

This continues as the basic distinction in discussing color sources. A pigment, no matter how finely it is powdered, will not dissolve in the liquid used to spread it. The tiny particles remain suspended in the liquid, and when the liquid dries, the particles join in a crust or coating.

Dyes do dissolve. The color becomes part of the solution or bath and this solution soaks into leather, cloth, or other fibers. When the material dries, the color remains.

Pigments are the key ingredients of paints and a great many inks. There are many other inks, however, which rely on dye for their color. Some inks use both pigments and dyes. Though it is possible to print or paint on cloth, no true dyes use pigments.

The difference between pigments and dyes, however, is not always clear-cut. Often we must consider the liquid used with them. There are some dyes which do not dissolve in water or oil. They can be ground up and used as pigments. They work as dyes only if special chemicals are added. There are pigments which will dissolve in some solutions such as acids but will work satisfactorily in water or oil. Lime, for instance, serves to make whitewash when stirred into water, but it will dissolve in acid.

With paints and inks, the liquid used to spread or carry the pigment is called the vehicle. With dyes, it is called the solution or bath. Vehicles rarely are simple mixtures, and preparation of a proper dye bath sometimes takes days.

The Egyptians and Hebrews, credited with the invention of whitewash, added milk curds to the mixture. Milk curds are a source of casein, a strong glue. It made the lime particles stick to each other and to the surface where the paint was spread. Today's paint-makers refer to such additives as binders.

The Egyptians are credited with discovery of mordants to improve dye baths. They found that certain metallic salts, even common table salt, made colors last longer in cloth. In a way, mordants too worked as binders. They made a chemical link between fibers of the cloth and colors of the dye. Alum, a salt combining potassium and aluminum, was one of the first mordants used. It remains popular today with craftsmen who use native dyes, but salts of tin, chrome, and iron serve as well. Sometimes a single dye will yield different colors with different mordants. Cochineal, a famous New World dye made from the bodies of insects, turns wool red when used with a salt of tin. With a chrome salt, however, the wool turns purple.

Dyes can be used to color pigments. Such pigments, called

A *dye worker of the 1800s winds cloth over and under a series of rollers, keeping the cloth moving continuously through the dye vat to assure even dyeing.*

SMITHSONIAN INSTITUTION PRESS

lakes, arose from the residue in dyers' vats. Colored fragments of cloth were collected, dried, ground up, and used to color paints. Later it was found that many white pigments would soak up and take on the color of a dye.

Of course, the early craftsmen had little concern for technical names for things such as vehicles, binders, mordants, and lakes. They simply adopted through trial and error those mixtures and methods which worked. As civilizations rose, skills slowly expanded and were passed from one generation to the next. Often such skills were retained as trade secrets within one family or community. Sometimes a region became famous for a special colored cloth, an ink, or a paint. The fame sparked trade, but it did little to expand man's knowledge. Actually, even without the shroud of secrecy, no one could explain why certain chemicals worked better than others or why certain complex methods produced the most lasting colors. The answers did not come until the scientific age dawned.

OUT OF DARKNESS

Aristotle, the famous Greek teacher and philosopher of the fourth century B.C., seriously delayed the study of light and color. He said colors were nothing more than various mixtures of black and white light. His reputation was so great that no one dared to challenge the theory with anything more than a whisper for nearly 2000 years. It was England's Sir Isaac Newton who put us on the right track. After extensive experiments with prisms, the triangular pieces of glass that bend light, Newton concluded that white light held all colors, blackness held none. Newton, living from 1642 to 1727, actually was not the first to challenge Aristotle. The Italian Renaissance painter and inventor Leonardo da Vinci (1452-1519) had concluded that color could not be seen without light, and England's chemist and physicist Robert Boyle (1627-1691) made a distinction between direct and indirect light. Da Vinci, however, was persecuted for his advanced ideas on light and other matters, and few of Boyle's fellow scientists understood the importance of his distinction.

Today, though our knowledge of light has advanced tremendously since the days of these pioneer thinkers, there are still things about it which puzzle us.

We do know that light is a form of energy—electromagnetic energy. Heat, radio waves, X rays, and gamma rays are some of the other forms of electromagnetic energy. All travel through space at the astounding speed of 186,000 miles a second. The forms of energy differ because they travel at different wavelengths, or pulses. The human eye is sensitive to a limited range of wavelengths—the portion of electromagnetic energy we call light.

Heat and radio waves are longer than light waves. We cannot see them. X rays and gamma rays have shorter waves. We cannot see them either. Of course, we speak of "waves" simply as a matter of convenience. Light does not move like the slowly rolling waves of the sea. It would be more precise to say that it vibrates with crests and troughs of strength that are so close together that only highly sensitive instruments can measure them. These tell us that the distances between crests of visible light waves range from .0004 to about .0008 of a millimeter. The waves of energy within these limits are called the visible spectrum.

We see different wavelengths of energy within this spectrum as different colors. Reds, for instance, have long wavelengths and are said to be at the "long end" of the spectrum. Blues and violets have short wavelengths and are said to be at the "short end" of the spectrum. When all waves of the visible spectrum reach our eyes together, we see white light.

If you have a prism, let a beam of light shine through it onto a white surface. Because the triangular glass bends light unevenly, it separates the wavelengths of energy and produces all the colors of the spectrum. Nature produces a more spectacular spectrum when raindrops bend and reflect sunlight to create a rainbow. Both the prism and the rainbow prove that white light is a mixture of all colors.

Of course, there is much more to the study of light than this. Scientists today have a theory that light is something more than waves of energy. They believe tiny units ("packets") of energy called photons are invloved. The photons, so small that they can pass through glass and other transparent solids, are activated by waves of electromagnetic energy to produce light. Albert Einstein (1879-1955) envisioned photons when he put forward his adaptation of Max Planck's quantum theory of energy. This explained a phenomenon of light known as the photoelectric effect—the ability of light to generate an electric current in certain metals. The electric eye which holds open doors of loading and unloading elevators and rings a bell in burglar alarms uses the photoelectric effect. Television would be impossible without this puzzling link between light and electricity.

For our purposes, we need not be overly concerned with photons, the quantum theory, and the photoelectric effect, but it is important to remember that our knowledge is not complete. The quantum theory remains just that—a theory.

KINDS OF LIGHT

In discussion of paints, inks, and dyes, we must consider four different kinds of light—direct, reflected, absorbed, and filtered light.

Direct light includes the energy that comes from the sun, a flame, and a light bulb. When direct light strikes an object, three things can happen. It can be reflected, absorbed, or it can pass through the object. What happens depends on the nature of the object. A mirror, for instance, reflects so much light that we can see our image in it. Most objects, however, absorb great quantities of light, reflecting only a select few waves of the visible spectrum. A leaf absorbs nearly all colors but green. The leaf thus reflects green. There are some reds and yellows reflected as well, but the green dominates. A ripe apple absorbs all colors but red. The red is reflected. That is what we see. The daffodil absorbs all but yellow. We see a yellow flower. An object's color is thus determined by what colors it absorbs.

There are some objects, however, which let most of the light pass through them. Glass is a classic example. Such objects are said to be transparent. There are still other objects which let only certain colors pass. These are called translucent. A drinking glass filled with milk seems to reflect all light. It certainy looks white. But if you hold it in front of a source of direct light, the milk appears pale yellow. Some yellow light passes through the milk. Stated another way, the milk filters out all light but pale yellow. Church windows with stained glass pass filtered light. From the outside, such windows seem dull. They absorb a great deal of the sunlight. From inside the church, the windows glow with filtered strong outside light. Of course, there are many objects which let no light pass at all. They are said to be opaque.

The four different kinds of light—direct, reflected, absorbed, and filtered—must be kept in mind when considering pigments and dyes. We must remember first of all that there are different kinds of direct light. Light from the bulb in a house fixture differs from the light from the tube in a fluorescent lamp. Both produce light quite different from sunlight. As a general rule, artificial light lacks many wavelengths, particularly those from the blue end of the spectrum. This is deceptive because artificial light and sunlight both appear white to the eye. But if you take a painting, cloth, or other colored object—preferably one with various hues of blue—

BELL TELEPHONE LABORATORIES

This photograph of two physicists in a horn-reflector antenna shows direct light in a patch on the floor and on their backs, while the fronts of their bodies and the walls are illuminated by indirect light.

from the sun into artificial light, it seems to change color. Some of the light waves reflected in sunlight are missing.

When mixing pigments or dyes, it is particularly important to remember they are basically absorbers of light. When we mix two pigments together, for instance a red and a blue, the result is a darker color. More light is absorbed, less reflected. This particular mixture gives purple. If we add yellow to it, we get an even darker mixture—almost a black. We have increased light absorption even more. Such mixing is called subtractive mixing because each new color reduces the reflected light.

Pigments and dyes can also work as filters. Light passes through them, hits the surface beneath, and bounces back to

Photographs of astronauts on the moon show harsh light effects because there is almost no atmosphere there, and no floating dust and moisture; if there were, the illumination would be softened by reflection of light from the floating particles and by their getting in the way between subject and camera.

NASA

the eye with some wavelengths removed. By controlling thickness of paints, artists can get either filtered or reflected light from the same pigment. Most watercolor paints are spread so thinly that they act as filters. There is a printing ink that appears red until it is applied thinly on paper. Then it is blue. Red is its reflected color. Blue is its filtered color.

Many of the lake pigments produce filtered light. The white pigment base acts as a reflector while the dye surrounding it acts as a filter. Actually, with lakes and translucent pigments, light is filtered twice, once as it goes in and again as it bounces back. Mixing filtered light is just the reverse of mixing reflected light. The filter, remember, removes all but selected wavelengths of light. Thus when two filtered colors are brought together, we get more light. A beam of red light thrown on the wall with a beam of green light produces bright yellow. This is called additive mixing.

When mixing colors, we speak about primaries. The primary colors for subtractive mixing are red, yellow, and blue. The primaries for additive mixing are red, green, and blue. Mix the subtractive primaries together and you get black. Mix the additive primaries together and you get white.

This sounds much simpler than it is. The trouble is that it does not always work. There is one great complicating factor—the human eye.

THE PUZZLE OF SIGHT

In 1959, Edwin H. Land, the American whose inventive mind led to the development of the Polaroid Land camera, made a remarkable experiment with cameras, filters, and projectors. Land took two pictures of the same scene, one through a red filter and the other through a blue filter. Then with two projectors, he cast the images of the two photos on a screen so they overlapped.

Since there was no green involved in the experiment, you would suppose that no green wavelengths of color would appear in the picture, but amazingly, all those who viewed the picture saw almost a full range of colors. Land made further experiments with other colors, using just two at a time, and got the same results. Somehow, the eye supplies a missing primary color. No one has yet explained the phenomenon.

There are other mysteries. Scientists have dissected the human eye and counted 18 million light-sensitive cells in the tissue. Yet there are just one million fibers in the optic nerve, the nerve that carries impulses from the cells to the brain. This suggests that there is selectivity—some kind of decision-making process within the eye—that sorts out messages of light before sending them to the brain. Does the eye "think"? No one knows.

The eye can be easily described as a camera. Light enters the eye through a lens at the front and is focused on a "film" at the back. The film, however, is far more complex than any ever devised for a camera. This layer of tissue in the back, called a retina, is made up of millions of two distinctly different kinds of cells—rods and cones. The rods are responsive to white light and dark and give us "night vision." In bright daylight, rods lose their effectiveness; the cones take over. Cones are sensitive to red, green, and blue light waves and mix them additively in much the same way we mix filtered lights, except that the mixing of the eye is reduced to a nervous impulse to the brain. No one knows exactly how this is done.

We do know that rods and cones have pigments in them that react to light and stimulate nerve fibers. Color blindness in certain persons is due to a lack of some of the pigments necessary to see a full spectrum of color. Usually, just one pigment is lacking, but in rare cases two are lacking. In such cases the eye sees a gray world.

Even those with normal eyes do not see colors alike. You

have probably argued with a friend over color. The friend may say a sweater is blue. You think it is green. The argument is hopeless. Another friend may like certain combinations of color, combinations that you find nauseating. Who can account for taste? The question is hardly scientific, but perhaps when more is known about how the eye sees, scientists might indeed account for taste. Psychologists have at least made a start—they know that if you dislike a certain color intensely, you have probably had an unpleasant experience in the past that involved that color.

ATOMS AND MOLECULES

Paints, inks, and dyes are chemicals, often very complex chemicals. Of course, we do not have to be chemists to use and appreciate colors, but understanding a few chemical facts can improve our use and appreciation a hundredfold.

In chemistry the smallest unit of matter is the atom. Atoms rarely occur in nature by themselves. They are usually joined with one or more atoms. Such combinations are called molecules.

There are many different kinds of atoms—over a hundred. There is a different atom for each element of the universe. In fact, an element is just that—a chemical composed of one specific kind of atom and no other. Oxygen, hydrogen, copper, iron, calcium, and carbon are elements.

When different atoms unite, we have a compound. When three, ten, twenty, or more atoms unite we still have a compound. With a hundred elements to begin with, and with unlimited numbers of combinations possible, we can see that the number of possible compounds is virtually unlimited. Chemists have a broad field of study indeed.

Compounds usually have characteristics quite different from those of their parent elements. Water, for instance, is a

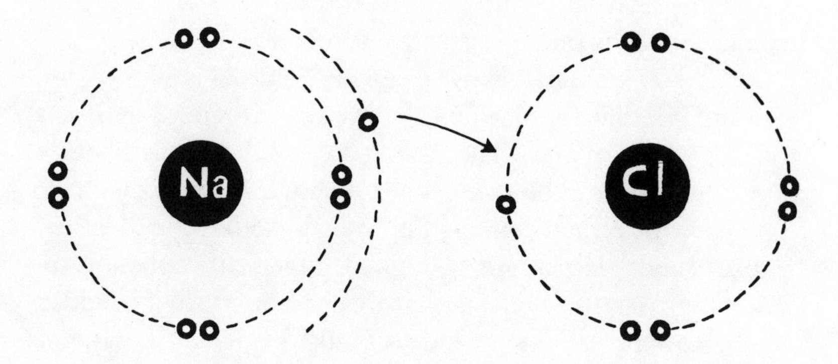

Common salt, which can be used as a dye mordant, is formed when an atom of sodium (Na) links its "extra" electron, in its outer orbit, with a chlorine (Cl) atom by filling the chlorine atom's "gap."

liquid composed of oxygen and hydrogen, both gaseous elements.

While the number of compounds is virtually unlimited, chemical groupings do follow set rules of nature. Perhaps the most important is the rule of valence. Not all atoms have the same power in linking with other atoms. Consider the creation of water from atoms. Hydrogen has the power for just one link. Its valence is always *one*. Oxygen, on the other hand, has the power for two links. Its valence in this combination is *two*. Thus when the two elements join to form water, it takes two atoms of hydrogen to one of oxygen to make a stable link. Chemists call water H_2O. The letters stand for the elements, and the numbers represent the number of atoms (over one) required of each element to form the specific compound.

Oxygen, of course, combines with many other atoms. Chemists call the process oxidation. When iron rusts, when wood burns, and when gasoline explodes, we have various examples of oxidation. Many paints and inks dry and harden through oxidation. Some dyes depend on oxidation for their color. Oxygen is thus a very active combiner.

Carbon is another active element. Charcoal is almost pure carbon. It is difficult to realize that we are surrounded by atoms of this black matter. The atoms are combined with other atoms to form carbon compounds. Every living thing is made up largely of carbon compounds. Wood, cloth, paper, leather, rubber, and plastics are composed largely of carbon compounds. Even some metals have carbon in them for added strength. Paints, inks, and dyes could hardly exist without carbon. Carbon molecules are often extremely complicated. Carbon can join with other elements in many different ways.

What makes carbon so versatile? For one thing, it has a valence of four, double that of oxygen. For another thing, it links with other carbon atoms as readily as it links with atoms of other elements. This means it can form huge molecules. Chemists speak of rings, chains, and lattices in describing the way carbon links with itself and other atoms. Sometimes they are simply called "giant molecules."

One of the surprising things these molecules can do is to change their structure without going through a chemical change. This involves a shift in linkage between the atoms. The shift alters the properties of the compound. Modern paints are some of the best examples. They can be stored as a liquid, but when they are spread thinly and exposed to air, the paint's compounds change to create a tough coating. Such shifting molecules are called polymers, and the shift itself is called polymerization.

Until well into the last century, chemists believed that nature alone could make the carbon molecule; it could not be made artificially in the laboratory. The notion was slow to die, even after pioneer chemists did indeed begin to make some of the simpler carbon compounds. In fact, the first man to do the feat turned away from his discovery, saying it was far too complicated. We will describe this and other discoveries in a later chapter.

Many of the discoveries have been in this century. They have brought us plastics, artificial textiles, a host of new paints, inks, and dyes, and many synthetic substitutes for such things as glass, rubber, and leather. And today's research chemists tell us exploration of the full range of possibilities has just begun.

In their work with the giant molecule, scientists discovered the direct link between chemistry and light. They found that a group of atoms—any number can be involved—will unite in such a way that they absorb certain wavelengths of light. These combinations of atoms are called chromogens. Synthetics chemists have learned how to create these chromogens in the laboratory, and the skill has been taken up by industry. We have advanced to the point where any color desired can be designed in the test tube. We have come to a bright world indeed.

To appreciate this world fully, to make sure we no longer take it for granted, let us look at paints, inks, and dyes separately, with a chapter on each. After each chapter, we will describe a few home projects that will make you even more familiar with colors and how to make them work and play for you.

2

Paints—for Everything, Everywhere

A trip to the paint store can leave you gaping. There are so many different kinds of paint today that even the shop owner has trouble keeping track of them all.

There are fire-retardant paints, fast-drying highway paints, reflecting paints, paints that glow in the dark, paints that prevent masonry and cement from cracking, nonskid paints for walks and stairs, paints for the insides of cans that prevent food from rotting, paints that stop rust, poisonous paints that keep barnacles from clinging to the bottom of boats and, of course, a vast array of artist's paints.

Paints can be rolled on, brushed on, dipped on, baked on, or sprayed on. You can even buy spray paints in push-button cans. In factories, there are paints that are given an electrical charge. They are designed to cover metals that carry an opposite electrical charge.

You may tell your shop owner that you want nothing more than an ordinary house paint. You will have to be more precise. The shop owner will ask if you want an interior or exterior paint, a water base, oil base, or synthetic base paint, a gloss, semi-gloss or non-gloss (flat) paint. If you are not certain of your needs, you may end up with the wrong covering. And then there is the question of color. The choice of color is so broad that your decision may take much thought. If you are not alone, it may also take much debate.

This situation is new. A hundred years ago the owner who could afford to paint his home had very little choice. Three hundred years ago artists could not find all the colors they needed to match what they saw in nature. Remarkably, they still managed to produce masterpieces. Thousands and thousands of years ago, back in prehistoric times, artists worked with no more than three or four reliable colors. The cave paintings, some of which are 50,000 years old, rival things you can see in today's modern-art museums.

We will have much to say about artists in this chapter. They searched constantly for new and better colors. They pioneered the development of paint. The first known artists are by far the most fascinating. In fact discovery of the cave paintings touched off a world-wide debate that lasted for twenty years. Few scholars were willing to credit stone-age men with artistic talent.

HALL OF THE BISON

The great debate began one November day in 1878. Nine-year-old Maria de Sautuola had accompanied her father, Don Marcelino de Sautuola, to a recently discovered cave near Altamira in Northern Spain. By digging in the floor of the cave, Don Marcelino hoped to add to his collection of spear points, scrapers, and other stone-age artifacts. Maria, not interested in digging, took a candle and began exploring the cave. She ducked into a low chamber where her candle flame revealed a breathtaking sight. Red, yellow, and black animals romped across the ceiling. Maria thought they were bulls. She began shouting, "*Toros! Toros! Toros!*" Actually, they turned out to be pictures of bison, an animal long extinct in Europe.

Don Marcelino came running, and soon he and Maria found pictures of other animals, including a running horse and a boar. They hurried home to announce their discovery

One of the bison pictures in the Altamira caves. It is interesting that this prehistoric painting has much in common with the style of some modern art.

to the world. No one believed them. Famous scholars came to examine the paintings and concluded they could be nothing more than well-prepared hoaxes. Don Marcelino, they said, was either a wonderful liar or a great fool. He died before the scholars changed their minds, but Maria lived to hear her father praised both for his wisdom and for his stubborn faith in the validity of the paintings. By then, many other cave paintings had been found in Spain, southern France, Sicily, southern Italy, and North Africa. There were pictures of giant cave bears and mammoths unknown to modern man. There were also pictures of deer, oxen, and wild horses. Charcoal and a few natural clays were used for pigments. No one can be sure what vehicle was used. It might have been water, animal grease, or raw egg.

The paintings evidently served religious purposes. Cave men lived by hunting. The animal paintings brought luck, guarded perhaps against avenging spirits, or assured regeneration of herds. Many of the animals depicted appear to be pregnant. The artists were the medicine men, or shamans, of the tribe. They knew their business. Scholars say that paint was applied with brushes of feathers or frayed sticks, with the fingers, with hollow bones which served as spray guns, and with the mouth. The artist would fill his cheeks with paint and squirt it on the walls. The cave artists may have lacked raw materials, but they had no shortage of ingenuity.

The artists who came after were often not as clever, but paints gradually improved, selection of materials increased, and painting methods broadened. Artists of Egypt, China, Babylon, Crete, and Rome all made their contributions.

We cannot attempt a history of art in this chapter. There are many fine books devoted to the subject, but the following descriptions of the main ingredients of paint will read at times like history. The advance of paint goes hand in hand with the advance of civilization.

The two basic ingredients—the pigment and the vehicle—will be discussed separately. Later we will take up special coatings such as varnishes, which have no pigments, and stains, which rely on dye rather than pigments for their color. But first let's consider the basics.

PIGMENTS

Nearly any solid which can be finely ground and will not dissolve can serve as a pigment. In 1849, when William Lewis Manly, who was to become the hero of Death Valley, led a boat expedition into the raging Green River of Utah Territory, he wrote his name on a canyon wall with a mixture of grease and gunpowder. His name remained clear and legible twenty years later when John Wesley Powell led the first men down the Green and on through the Grand Canyon of the Colorado.

Gunpowder, of course, makes a rather expensive pigment. It is not recommended.

Makers of paint pigments divide them into three groups—

white pigments, colored pigments, and extender pigments. Here are important examples of each group.

White Pigments—White lead has been found in the oldest known paintings of China. It was certainly one of the first manufactured pigments. Lead carbonate, as the chemists call it, is made by corroding raw lead in vapors of acid. The Greek scholar Theophrastus (372-287 B.C.) wrote of artists making white lead, but mass production of the pigment did not get under way until the sixteenth century, when the Dutch developed a practical system to produce acid vapors by fermenting vegetable matter. Though faster methods have since been developed, the so-called "Dutch process" is still used today.

White lead is extremely durable. Many old paintings have been found in which all but the white lead has decayed. Artists call finely-ground white lead "flake white," and because it mixes well with oil, artists used it for centuries as a standard in gauging other pigments. With water white lead is a poor mixer, but that is not its worst fault. White lead can kill. Swallowing it causes lead poisoning. Also, sulfur fumes such as those found in industrial smoke turn white lead dirty brown, even black. For this reason, use of white lead has declined in favor of other pigments.

Zinc oxide is made by vaporizing zinc ore in an ample supply of oxygen. The French began making it about 200 years ago for artists. An English firm offered a watercolor with zinc oxide by 1834. Particles of zinc oxide are not as nearly opaque as white lead. It thus lacks the same covering power. This means that two coats of zinc oxide are sometimes needed to do the job of one coat of white lead. Zinc oxide does mix well with water, but it tends to crack or flake with age. Some paint-makers combine zinc oxide with white lead for a more durable paint. Artists use a great deal of zinc oxide. They call it zinc white.

Lithopone, a mixture of zinc sulfide and barium sulfate, has limited paint use today, mostly in undercoatings. It has high covering power, but the chemicals involved deteriorate with time. Thus lithopone, used extensively in interior house paints until the 1950s, has largely been replaced by newer whites.

Titanium dioxide, the leader of these new whites, appeared in commercial quantities in the United States and Norway in 1919. Production has grown steadily. A lightweight pigment with high covering power, it resists other chemicals and heat better than almost any other pigment. In addition, titanium is the world's ninth most abundant element. We are not likely to run out of it. Titanium dioxide does turn soft

A worker adds a sack of titanium dioxide to a high-speed dispersion tank in a modern paint factory. The use of this pigment has grown steadily since 1919.

GLIDDEN-DURKEE DIV. OF SCM CORP.

or chalky when it is used with oil. Thus other pigments are usually mixed with it in oil-base paints. But because it stands heat up to 1500° Fahrenheit, titanium dioxide is used extensively in paints for stoves and other kitchen appliances.

Colored Pigments—Customarily, colored pigments fall into two divisions, organic and inorganic. Organic simply means that the chemical contains carbon. Thus organic pigments are those that come from living matter or from the synthetics chemist's laboratory. The inorganic pigments come from mineral sources. Another distinction is sometimes made between native and man-made pigments. Here we will simply group the pigments by their colors. For convenience, black will be included here, though in a strict scientific sense it is not considered a color.

Blacks: Charcoal, soot, lampblack, and other nearly pure carbons produced by incomplete combustion make excellent black pigments. The Chinese began making lampblack by burning oil in small pots, but there were many earlier sources of carbon. Vine cuttings, fruit pits, bones, and ivory produced rich blacks when burned. Tones ranged from brownish black to jet, depending on the amount of impurities. Most of today's black comes from incomplete burning of coal and natural gas. Burning of gas to produce what is known as carbon black began in the United States in 1864. In nature, black oxides of iron, manganese, and cobalt have served artists from ancient times. Manganese oxides dug from natural deposits were used in a few of the cave paintings in place of charcoal. Asphalt, or pitch, still used as a roof coating today, has been found on mummy cases in Egyptian tombs. This asphalt was gathered from deposits near the shores of the Dead Sea.

Reds: Red ocher, one of the natural clays, came into use far back in prehistoric times. Powdered ocher was used in stone-age burials. Red and yellow ochers are the main pigments found in the famous stone-age paintings. The ochers continue to be widely used today. It is impossible to say exactly when artists found that roasting these clays would change their color. Yellow ocher turns red with heat. Tans, such as siennas and umbers, darken and take on red hues under heat. The iron oxides which give these natural clays their color can be produced chemically. These inorganic, man-made pigments, called Mars colors, are available in many shades.

Another early red came from cinnabar, the ore of mercury. Vermilion, as this pigment is known, was used in a stamping ink by Chinese artists. The red stamp in the corner of paintings served as the artist's signature. Vermilion has also been found in ruins of Assyrian houses. The Bible cries woe on the unrighteous who paint their homes with vermilion. Greek traders gained wealth by shipping cinnabar from rich mines in Spain, but like the iron oxides, vermilion was eventually made artificially. European craftsmen learned to make it in the eighth century by combining mercury and sulfur.

Some of the first lake pigments used by artists were reds. Madder, kermes, and Brazil wood, which we will discuss later, supplied the dyes for these lakes which served thirteenth-century painters. Cadmium red, another artificial or man-made chemical compound, was first produced in Germany in 1907. Cadmium red remains today as an important art and commercial pigment.

Yellows: Many metallic compounds make yellow. Zinc yellow, cadmium yellow and chrome yellow are among the most important yellow pigments. Lead antimonate is a man-made compound artists call Naples yellow. Traces of it have been found in bricks of 2500-year-old ruins of Babylon. Strontium and barium compounds also make yellow pigments, but for a long time the most reliable and sometimes the only yellow available was ocher. Like red ocher, the yellow gets its color from iron oxide, but a slightly different form of it. This is why roasting yellow orcher changes its color.

Early artists of Egypt, Greece, and Rome used yellow ocher extensively. Today's fine artists prefer a yellow ocher mined from clay deposits in France. They insist that no other yellow can match it. Chrome yellow and lead chromate—both man-made compounds—are the most common yellows in house and industrial paints, however.

Blues: Finding a blue pigment challenged the ingenuity of the ancients. Lapis lazuli, the semiprecious gem found in Iran, Afghanistan, and China, could be ground up and cleaned for use as a pigment, but it was very expensive. As early as 3000 B.C., Egyptians found a substitute by melting a mixture of sand and salts of copper and other metals. The pigment, called frit and later Egyptian blue, has been found on wall paintings in the ancient palaces of Crete. The Minoan artists who painted them either imported frit from their Egyptian neighbors or learned to make it for themselves.

Indigo, most often used in dissolved state as a dye, was sometimes used as a pigment. Indigo would not dissolve without special chemicals, and paint-makers took advantage of this. Prussian blue, a combination of potassium or ammonium salts and iron cyanide, made its first appearance in a Berlin laboratory in 1710. Cobalt blue, a combination of cobalt oxide, aluminum oxide, and phosphoric acid, appeared in France in 1802. Manufacture of artificial lapis lazuli began in France in 1828. It was made by melting clay, soda, and sulfur in a coal furnace; it was called ultramarine.

Greens: Malachite, a native copper compound known to the ancients, had a yellowish tint. Egyptian women reportedly used this green for eye shadow. Verdigris, another copper compound, had a bluer tone. Wall paintings found in the buried city of Pompeii, Italy, show that verdigris was a popular color among artists in the first century A.D. The compound is poisonous, but substitute greens were slow to develop. Emerald green, even more poisonous, was discovered in Sweden in 1788

NATIONAL PAINT AND COATINGS ASSN.

To obtain green household paints, as for painting shutters, most manufacturers today mix iron blues with chrome yellows.

and first produced commercially in Austria in 1814. It is a chemical combination of copper and arsenic. It was 1838 before another green, viridian, made with chromium, appeared in Paris. Artists could use it in safety. Most green in commercial paint today comes from mixing chrome yellows with iron blues. In addition, a heat- and weather-resistant green is made from chromium oxide.

Browns: A natural clay known as umber was used by the ancients. As we have seen, it could be darkened and given a red tint through roasting. Today the color can be made artificially by combining iron and manganese oxides. Other browns of broad color range can be made by subtractive mixing of yellow, blue, and red pigments.

Special pigments: Tiny flakes of aluminum, copper, or

bronze serve as pigments in silver and gold paints. Gold leaf only rarely uses real gold; it is most often a combination of copper and bronze. Zinc and lead flakes have fine antirust properties for metallic paints, and zinc and lead compounds are used in protective coatings for bridges, ships, and other large structures that would otherwise be destroyed by rust. Luminous chemicals make paints that glow in the dark—by themselves or under ultraviolet light. Some of these are particularly useful in highway signs and markers.

Lakes: The vast number of organic, man-made dyes—the synthetics—provide the paint industry with a full spectrum of colors. Any of the old, inorganic colors can be matched by a synthetic dye lake. In addition, there are colors available now that were never available before the advent of the synthetics industry. As a general rule, however, the lake pigments do not withstand weathering and sunlight as well as most of the old pigments. Lakes thus have their widest use in artist colors and interior house paints. Alizarin, the dye of the madder plant, first synthesized in 1870, and toluidine and lithole pyrasoline, are used in making red dye lakes. Phthalocyanine (pronounced thalo-cyanine) is a versatile dye that is used to make blue and green lake pigments.

Extender Pigments—Extenders, sometimes called inert pigments, are chosen not for their color but for their consistency. They improve storing qualities, ease of application, and durability. In addition, extenders are generally cheaper than colored pigments. Their use saves us money when we buy paints. Here are some of the more important extenders.

Calcium carbonate—Limestone, chalk, whiting, and powdered marble all come under this heading. Calcium carbonate resists abrasion and weather well. It tends to turn muddy in oils, but it retains its white color in water-base paints. As the various names for it suggest, calcium carbonate occurs in various crystalline structures. Whiting, with good binding qualities, is mixed with oil to create the paste we call putty.

Barium sulfate—This is a transparent compound that artists call *blanc fixe*. It resists other chemicals and stands up well under abrasion. It is often added to paints to prolong the life of the finished coating and is particularly useful in exterior paints. Barium sulfate also serves as the base for many lake pigments.

Talc—Chemically known as magnesium silicate, talc occurs in natural deposits throughout the world. It has a slippery texture which improves brushing quality of paints, and when mixed with other pigments it helps to keep them from settling. This is particularly useful in paints that must be stored for several months. In addition, talc contributes to durability.

China clay—Another native compound, aluminum silicate to the chemists, China clay reduces settling and serves as a lake base. Its use is somewhat restricted, however, because it tends to muddy some colors.

Mica—This flaky, transparent element toughens coatings and protects colors from fading in the sun.

Aluminum hydrate—This compound is used in nearly all paints packaged in tubes. It provides the necessary buttery consistency for such paints and retards settling and drying. It is transparent and thus serves as a lake base, particularly for brilliant colors that would be toned down by a white lake base.

VEHICLES

While pigments determine the color and have some influence on the durability, storage qualities, and other traits of paint, the vehicle selected to spread the pigments determines how the paint behaves. In fact most paints are described by their vehicles. Thus we have such terms as water-base, oil-base, tempera, lacquer, and enamel paints. All these terms describe the vehicle involved.

Vehicles must meet several requirements. They must be easily obtained. They must mix easily. They must not react with or change the color of pigments. They must not dry rapidly when stored, but they have to dry rapidly when spread in a thin coating.

Water, it would seem, meets all these conditions perfectly. It does except for one fault. Water alone has no binding properties. It dries out completely, leaving pigments in their original powdery state. Hebrews and Egyptians, as we have seen, added milk curds to bind the lime pigments of whitewash. This was not only the first use of whitewash, but also the first use of casein, the natural glue of milk, as a binder. Casein paints are widely used today among artists and industry despite the recent introduction of many synthetic, water-soluble binders. The Chinese developed a water-base paint probably well back in prehistoric times by using gum arabic as a binder. Gum arabic is a water-soluble gum secreted by acacia trees found in Africa, Australia, and Asia. The Chinese ground both pigment and gum together, added enough water to make a paste, and then dried the paste in cakes. When it was ready to use, the cake was ground on a small, stone pallet with enough water added to meet the artist's requirements. The Chinese made their ink the same way, although many other ingredients were added to the gum and pigment. Gum arabic continues as a popular binder, particularly in artists' watercolors.

Casein, gum arabic, and other binders make what are generally classed as tempera paints. One of the earliest tempera paints, perhaps earlier than casein or gum arabic, was a water-base paint that used egg as a binder. Cave artists may have used the white, the yolk, or the whole egg in mixing their paints. Certainly the Egyptian artists mixed paint with egg.

An ancient Greek artist painting a statue in encaustic. Redrawn from the decoration on an early Greek vase.

Artists who painted common scenes of Egyptian life on tomb walls to comfort the spirit in the hereafter may have used gum arabic and egg tempera paints interchangeably.

The Egyptians painted on dry walls of mud or plaster. Minoan artists on Crete were the first we know of to develop wet-wall, or fresco, painting. In effect, wet-wall paints use the drying plaster itself as binder. The pigments, mixed with water, soak into the plaster before it dries. When drying does occur, the color remains, a vibrant and durable part of the wall. Many of the Italian Renaissance artists won their reputations with frescoes. Michelangelo's paintings in the Sistine Chapel remain today as leading examples of the technique. Mexican artists have carried fresco painting into modern times.

Another paint vehicle discovered by the ancients was hot beeswax. Hot-wax painting, known today as encaustic painting, was practiced by Egyptians, but the Greeks evidently brought it to a high art. We have to say evidently because no samples of Greek wax painting have survived. We know it only through the writings of Greece's classical era (480-300 B.C.). There may have been another vehicle in addition to wax, but we know even less about that. Some scholars suggest that the Greeks invented oil painting. Others say they found a way to dissolve wax in water and thus saved the trouble of heating their paints.

Heating was a problem. It was not only necessary to keep the wax heated to a liquid state, but it was also necessary to use hot irons or pokers to make corrections. Wax, however, imparts a fine gloss and shows pigments off with brilliant clarity. Encaustic painting continued into the Renaissance until it was replaced by easier, more versatile oil painting. Few modern painters have experimented with wax painting. One exception was the late Diego Rivera, most famous for his frescos, who did a large mural in wax on a hotel lobby wall in Mexico City.

The origin of oil paints is unclear. Oil as a paint vehicle was first mentioned specifically by Aetius, a Greek physician of the sixth century A.D. He does not make clear if this was something new or something artists had known of for centuries. European artists were making limited use of oils by 1100 A.D., a time when most work was being done with egg tempera, fresco, or encaustic paints. Art historians at one time credited Jan and Hubert van Eyck, Flemish brothers who painted in the early fifteenth century, with the discovery of oil paints. Today, historians agree that the Van Eycks simply popularized the technique. They were skillful artists and they demonstrated that oils could expand the means of expression far beyond traditional paints. Spread thinly, oils made pigments work as transluscent filters for light reflected from the canvas beneath. Spread thickly, they could add texture to the picture surface. In addition, oils were both pliant and durable. It was possible to roll up a canvas painting after the paints dried, allowing easy storage and shipment.

Of course, one had to be careful in selecting oils. Only drying oils, those that oxidized when exposed to air, worked with paints. Mineral oils would not dry, and some vegetable oils did not dry fast enough. There were many drying oils to choose from. Fish oil, particularly from the menhaden or pilchard, walnut oil, and poppy seed oil all worked well. Tung oil from a tree that grew in China did not become readily available until paint-makers imported the tree and began cultivating it. Today, linseed oil extracted from flax seed is by far the most popular in both commercial and art oil paints. Its drying powers can be improved by boiling or sun baking.

Pigments vary in their effect upon the drying rate of oil. Umber, for instance, hastens the rate while vermilion slows it. Some extenders speed drying. In addition, there are special chemicals which improve the rate. Some oil paints have to be thickened to assure proper drying, thickened so much that they do not brush easily. In these cases a liquid that evaporates readily in air is used to thin the paint. Turpentine, distilled from tree sap, makes an excellent oil paint thinner. Though it does leave an odor, turpentine evaporates rapidly and does not create a fire hazard. Some of the modern spirits refined from petroleum, while odor-free, cannot be used near an open flame. In a closed room, their fumes can damage a painter's lungs and even cause fainting and nausea.

VARNISH, LACQUER, AND STAIN

There is a group of unique coatings that can be used without pigments. All of these, the varnishes, the lacquers, and the stains, can also serve as vehicles for pigments.

Varnish is a resin, either natural or synthetic, that is dissolved in oil, alcohol, or some other solvent. Natural resin comes from the sap of trees and can be collected either in a fresh, sticky stage after it hardens or in a fossilized state. Fossil resin is dug from the trees of long-buried forests. The world supply has dwindled. The synthetic resins are generally tougher than the natural resin, but for clear coatings on fine furniture and hardwood floors, a natural resin varnish is still preferred for appearance' sake. It does not have the plastic-like gloss of a synthetic resin varnish. Artists sometimes use a clear varnish as a protective shield for a finished painting. Art shops now offer a varnish in a push-button spray can.

Shellac is a resin used to make a special varnish. The lac insect which lives on the sap of trees produces shellac. Alcohol usually serves as the thinner. Thus a coating of shellac dries quickly. It is ideal for filling in pores in wood and sealing against moisture, but it has some faults. Shellac scratches easily and darkens with age. Today it is used mostly as an undercoating in wood finishing.

Varnishes mixed with pigments make paints we call enamels. They provide tough, waterproof coatings with a glossy shine. Many of our autos are painted with enamels that use a synthetic resin varnish as the vehicle. Porch furniture, window frames, and the cabinets and walls in bathrooms and kitchens are usually painted with enamel.

Lacquers are produced today by dissolving cotton in various acids to make a clear solution of nitrocellulose. Chemically, lacquer is closely related to gunpowder and celluloid. There is also a natural lacquer exuded from a sumac tree of Asia. This was first used by Chinese craftsmen as early as 1027 B.C., according to ancient chronicles. The Chinese applied coat after coat on wooden trays, bowls, boxes, and other wares to obtain a high gloss. Special solutions such as acetone, ethyl acetate, and butyl alcohol are required to thin lacquers. These have the advantage of drying quickly, but they are also flammable. Thus care must be taken in storing and using lacquers. With pigments, lacquers make a glossy, fast-drying paint. Fingernail polish is a familiar example of modern lacquer.

Stains use penetrating oils as the vehicle and dyes or pigments for color. Wood-finishing stains are designed to darken wood to match the appearance of various hardwoods. Walnut, maple, oak, mahogany, and many other stains can be found in the paint store. Such stains can be applied with a brush or cloth. Excess stain is usually rubbed off soon after application. The longer this step is delayed, the darker the wood will become. Stains do not protect the wood. They are usually followed with one or more coats of varnish.

Stains containing pigments are relatively new, but they are popular today as exterior house paints, particularly for houses with sidings of rough wood. The penetrating oils carry the pigments into the wood. Such stains leave no brittle coating that might chip and blister in the sun or wet weather. A stained house, therefore, does not require repainting as often as one covered with conventional paint. Little wonder stains are popular with homeowners.

NEW PAINTS

While water-base, resin, and lacquer paints dry by evaporation and oil-base paints by oxidation, many of the new paints dry by polymerization. As we have seen, this is a change in the molecular structure. Liquids become tough solids without any chemical change. Sometimes a chemical called a catalyst is needed to bring about the change. Such paints are often called "two-part" paints. Usually, the molecular change in polymerizing paints cannot occur below certain temperatures. Using them outside on a cold winter day may not be possible.

The polymerizing paints or polymers fall into two classes: those soluble in water, such as the popular latex paints, and those that require special thinners, such as the alkyd paints.

Among the latex paints, one of the most popular is the rubber-base paint. It was developed first as an interior paint, but today there are also rubber-base exterior paints. The chemical groups involved in this synthetic are the same as those used in making artificial rubber; hence the name. These paints are practically odor-free and dry to a tough, washable surface as good as or better than that of any oil paint; but the best feature is their water-solubility. No special thinners are needed, and brushes or rollers can be cleaned easily in soap and water.

Acrylic and vinyl paints, also in the latex class, are currently very popular with artists. These are allied closely with plastics. Acrylics use the same chemicals needed to synthesize Plexiglas and Lucite. To the chemist, there is little difference between a can of acrylic paint and a pair of false teeth or a nylon stocking. Vinyls rely on the same chemicals used in making plastic table tops, kitchen counters, and artificial adhesives. Acrylic and vinyl paints can be regarded as enamels since they are based on water-soluble resins, but they do not all give the appearance of conventional enamels. Many of the house paints give a nonglossy finish, and the art paints range

A paint researcher controls a winch that lowers a tire rim into a special coating being tested for electrocoating use. In electrocoating, the paint and the object to be covered have opposite electrical charges.

GLIDDEN-DURKEE DIV. OF SCM CORP.

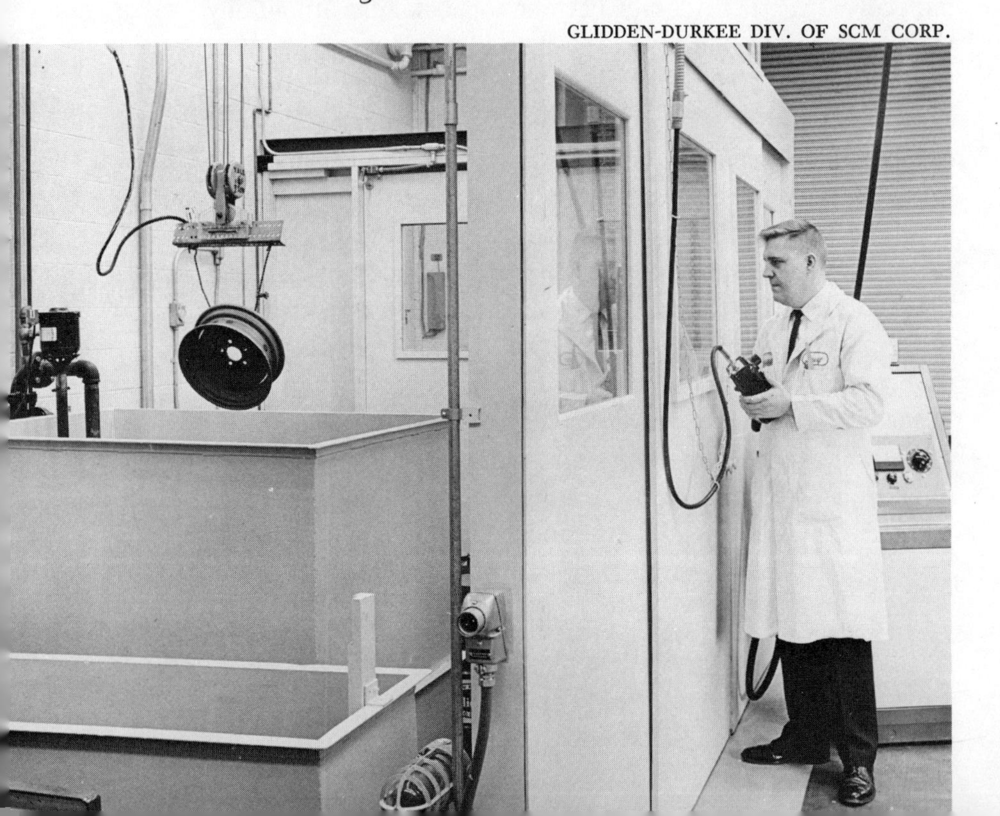

from imitations of water colors and oils to finishes entirely different from any conventional paint. No matter how they are made to appear, the acrylics and vinyls are far more durable than anything used by artists of the past. This accounts in large part for their popularity.

The second large class of polymer paints—the alkyds—use complex combinations of acids and alcohol. The name, in fact, rises from an attempt to combine the two words. These, too, are synthetic-resin paints, just as durable as the others, but the need for special liquids to thin them and to wash up after work has limited their use among home painters and craftsmen.

The trade names for many of the new paints can be very confusing. Such names as "acrylate," "polyvinyl acetate," and "epoxy resin" mean more to the chemist than they do to the painter. Be sure to read the instructions on the can before buying these fancy-name paints. Some are designed for highly specialized uses, and many require special thinners.

MAKING PAINTS

Despite the many synthetics available today for new paint vehicles, paint-making still begins with grinding. The pigments must be powdered. The process has been going on for thousands of years. Today motor-driven mills grind paint pigments, but not so long ago the job was done with stone mills driven by hand. A stone trough with a round grinding stone that could be rolled back and forth was typical of the early pigment mill.

In colonial America, the virgin forests provided a rich supply of turpentine and other resins needed to keep the wooden ships of the British navy in service. The tree gum products helped preserve both spars and planks of the ships, and the ships, ironically, often served to keep the rebellious colonies in line. Such products, as we have seen, are not actually paints.

Paint production in the colonies, as far as we know, did not begin until 1700, when Thomas Child arrived in Boston. He brought with him from England a stone trough and ball for grinding pigments. Child died six years later with little to show for his paint-making venture. The trough and ball ended up in the wall of a house, and they can be seen today in the old section of the city. In 1772, records show that the Mordecai Lewis Company was importing white lead, red lead, vermilion, and chalk from England. The firm was evidently the first successful paint company in the New World. Pigment-making did not start until after the Revolution. Samuel Wetherill began making white lead in Philadelphia in 1804. This was to meet the growing demands of artists and carriage painters for pigments ground in oil. White lead served as a base pigment. Various other pigments already thoroughly ground in oil served as tinting colors. For many years up into this century, the mixing system continued—oil, white lead, and tint—and the user put them together himself.

Our modern paints with all ingredients premixed and packaged in a single container, were first offered by the John Lucas Company of Gibbsboro, New Jersey, in 1860. Many other firms began making premixed paints, but they were not always successful. Often the oils and pigments separated during storage so that painters had to stir them just as long and hard as if they were mixing their own paints. New preparations and methods were needed, and these did not develop until the start of this century, when paint-making became a science.

Today most paint factories are housed in two- or three-story buildings. The reason is gravity. The pasty ingredients flow from top to bottom as they go through the various stages of mixing, grinding, testing, and packaging. On the top floor, the pigments, the extenders, and other ingredients are mixed with enough vehicle to make a paste. Measurements are exact, in keeping with chemically prepared formulas. Mixing usually is done in a tank with motor-driven paddles. From the tank the paste flows down through large pipes to the grinding room below.

Designs vary, but most paint grinders consist of three to five rollers that turn against each other, crushing the pigment particles and dispersing them evenly throughout the paste. When one of these roller mills, or grinders, has done its work, the paste has the consistency of butter. From the grinding room the paste flows again by gravity to the blending room.

Roller mills have long been among the most efficient devices for thorough and even grinding of paint pigments.

NATIONAL PAINT AND COATINGS ASSN.

Thinners, tinting colors, if any, drying agents, and any other paint additives required are placed in a tank with the paste. Again, motor-driven paddles mix the concoction. Testing follows. There are hundreds of ways to test a paint, but most of the factory tests check on color uniformity and balance of ingredients between one batch and the next. The tests of sun and weather resistance, covering power, and storage power were all made when the paint was first designed. Factory tests are made to assure that each new batch is in keeping with the original design.

Ball mills are efficient grinders for tough pigments. The worker at the top is opening a valve to deliver a new batch of pigment for grinding.

GLIDDEN-DURKEE DIV. OF SCM CORP.

Instead of roller mills, some plants have ball mills for grinding paint. A ball mill is a large, steel drum that revolves on a horizontal axis. It is loaded with many tons of steel or porcelain balls. As the drum turns, the tumbling balls grind and disperse the pigment. Ball mills can work as blenders and thus often serve to simplify the steps in paint-making. The paste that flows from the ball mill sometimes needs the addition of only a thinner before being packaged in cans or barrels.

Still another paint mill, known as the speed disperser, works with some pigments. It consists of high-powered blades whirling inside a sturdy tank. This is faster than the roller or ball mills, but it is not always as thorough.

With many pigments, straining is a necessary step in paint production. When thinned enough, the paint can be forced through finely meshed screens to remove foreign particles. In some cases centrifugal separators can be used. The heavier foreign particles are carried to the outside of spinning tanks.

Modern paint-making requires complicated and expensive equipment, but today's paint buyer should not want it any other way. The time and work involved in grinding and mixing paints by hand—the old method—would price paints beyond the reach of most buyers. You will appreciate this fact more after trying some of the projects described in the next chapter.

3
Paint Projects

The preceding chapter has undoubtedly given you some ideas of your own on making paints. Try your ideas. This is the best way to learn about paints, and the only way, incidentally, to discover and invent. The few projects described in this chapter should be considered no more than guides for further experimentation on your own.

Let's begin with making pigments. This takes grinding. Art shops sell mortar and pestle sets for grinding pigments, but you can get by with an old cooking pot and the handle of a hammer or other hand tool.

BRICK

The iron oxide of red ocher can be had in abundance from a common brick. The mural artists of Crete crushed bits of red pottery to make a pigment they called terra cotta, nothing more than iron oxide. You can try grinding up a broken flower pot if you wish, but a brick is easier to work. The results are the same.

First break off small chips by hammering at the edges and corners of the brick. Wear goggles or dark glasses to protect your eyes from flying grit. Next collect the chips and put them in your pot or mortar. Add a small amount of water and begin pounding. Continue until you have an even paste. While

Grinding brick for a home-made red pigment is a slow process, but a small amount of water added will soften the brick and speed up the work.

pounding, you may find that the paste sticks to the hammer handle. Slight additions of water may be needed.

Best binders for this pigment are egg yolk or glue. One egg yolk is needed for two tablesoons of paste. With a fork stir the egg yolk in a separate container with one-fourth cup of water. It is important to stir this mixture thoroughly before adding it to the paste. As you stir the paste and yolk mixture together, add more water gradually until you have a free-flowing, brushable mixture. Final stirring can be done with a stiff brush so you can test as you stir. You have made an egg tempera paint, one of the oldest known to man.

If you use glue, be sure you select one that is water-soluble. The white glues, though they lighten the color of the paint slightly, give the best results. One teaspoon of glue is needed for two tablespoons of paste. The glue must be mixed thoroughly in a separate container with one-fourth cup of water before it is added to the paste. Again, stir and add more water until you achieve the desired consistency. This is another tempera paint.

Both, of course, are water-base paints with water-soluble binders. Both dry with a gloss and rich display of color.

You can also make an oil paint with brick chips, but expect difficulties. Your pot and hammer handle are not the best tools for working with oil because of its sticky consistency. Turpentine will thin the mixture as you grind but expect to pound three to four times as long as you did with the water-base paints. Use boiled linseed oil. Thickness varies, but you will need about a quarter of a cup to make a brushable paint with two tablespoons of brick chips.

CHALK

While grinding brick chips takes a good deal of labor, chalk grinding is fast and easy. Use ordinary blackboard chalk. One stick of chalk, you will find, produces surprisingly little powdered pigment. This is because chalk is porous and contains a good deal of air.

Grind and mix chalk in the same way you did brick. Two sticks of chalk should be used with every egg yolk or with every tablespoon of glue. You can also use liquid starch as a binder with chalk. Use two tablespoons of starch for every

An egg yolk, thoroughly stirred with a small amount of water, serves as one of the best binders in home-made paints. The yellow color of the yolk fades when the paint dries.

two sticks of ground chalk. Chalk tends to take on colors from vehicles. The egg yolk will give it a yellowish tint, but normally that soon fades after drying. Oils turn chalk muddy, however, and we do not recommend attempting an oil-base paint with chalk.

You can make a lake pigment with your chalk. Grind two sticks of chalk with a teaspoon of liquid, all-purpose fabric dye, available in most markets. Dark reds and blues work best in this project. No dye will completely overcome the white of the chalk. You will get light reds and blues. Mix your paint as before with egg, glue, or starch as the binder. Your paint will finish to a high gloss, but do not expect the color to hold up long in direct sunlight.

Precolored chalk is available in green, red, yellow, and blue. These provide a shortcut in making a lake pigment. Simply grind and mix these as you would white chalk but add water slowly. The consistency of these colored chalks differs. Some will take more water and some less than others to attain the right paint consistency.

Pastels are chalk sticks which have been treated with a binder. By grinding a set of pastels, you can produce a full range of colors much more brilliant than you can make with dyed or colored chalk. Pastels, however, are expensive. Experiment with chalk first.

CHARCOAL

You can make a high-quality lampblack by holding a water-filled glass or bottle over a candle flame. Keep the glass or bottle moving so that it does not overheat and break. Unfortunately, this process is very slow. For home projects you will need much greater volumes than can be produced from a candle flame. Sticks of burnt willow sold by art shops for charcoal drawing are the easiest source of carbon. Lumps of charcoal from the fireplace will work, but the impurities in them will produce browns and grays.

Grind charcoal dry. It absorbs so much moisture that binders or water turn it to an unworkable paste. Add your vehicle after the willow sticks have been ground thoroughly. You will also find that absorption varies from one batch of ground charcoal to the next. Consequently, it is not possible to give a fixed recipe, but you can start with the same proportions you used in making paint from brick dust. Two tablespoons of powdered charcoal will require one egg yolk or one teaspoon of glue. Add more water slowly until you get a workable consistency. A glue binder will give the blackest, glossiest paint. Egg yolk tends to give a gray tone.

Wear old clothes for this project. Charcoal has a way of getting into places where you don't want it.

Boiled linseed oil can be used to make a black oil paint. The charcoal mixes with oil much better than brick dust, but even so, you may have to use turpentine to thin the mixture. The paint will not be quite as black as the one made with glue binder.

PREPARED POWDERS

Art and stationery shops carry canisters of powdered tempera paint that can be simply mixed with water and used. The powders contain both pigment and binder and are widely used in schools because of their ease of storage and handling. In fact at the end of the school year, teachers often discard partially used canisters when they clean up their shelves. You might be able to obtain some of this paint for nothing, but even if you have to buy it, the powdered paint is the cheapest you can find without grinding pigments yourself.

Addition of glue mixed to a syrupy texture will give the paint a hard, enamel-like finish. You can also use egg or starch as an additional binder. Remember, however, the powder already has a binder in it. Adding too much may produce a surface that crusts or flakes as it dries. Experiment with small batches to get the right proportions.

The powdered paints are ideal for dry-wall murals. A coating of clear varnish over the finished painting will protect the colors from fading. Verathane, a synthetic varnish, will add luster to the colors.

The powdered paints also lend themselves to experiments

with texture painting. You can add sand, grit, coffee grounds, even small scraps of paper to the mixture to give your work a rough, three-dimensional surface.

WAX

Common crayons make handy tools for experiments in encaustic painting. You can draw a "painting" by using a candle flame to soften the tip of the crayon before each stroke on the paper. This is an effective but slow process. If you want to cover a large area, you will need hot wax paint.

Mix crayons with paraffin on a one-to-one proportion. The paraffin provides brushability and guards against cracking when the wax dries.

Since overheated wax will catch fire, never put a pan or can of wax directly on a flame or electric heating element. Use

The home craftsman can imitate encaustic painting by melting the tip of a crayon as he works. Several different colors have been used in painting this fish.

the double-boiler technique. Tin cans of wax in a pan of simmering water make a safe arrangement. Paraffin melts between 117° and 149° F. This means you need not bring the water to a boil. Of course, a regular double boiler is more efficient, but it is far easier to discard cans at the end of this project than it is to clean hard wax residue from kitchenware.

Before starting you must of course remove the paper wrappings from the crayons. Do not be afraid to mix colors. A red and a yellow crayon melted together give orange. You can make green with blue and yellow, purple with red and blue, and so on.

Use old brushes for your painting. You can remove most of the wax with hot water when changing colors and make a final cleaning with turpentine and soap at the end of the project, but this treatment is rough on brushes and may spoil them for other uses.

Paint on heavy-duty paper or cardboard. We advise tacking this to a drawing board to prevent curling as the wax hardens.

As an extra precaution when working with hot wax, it is a good idea to have an open box of baking soda handy. If the wax should catch fire, a liberal shake of soda will snuff the flame.

Crayon does not mix with water. Try filling in the untouched portions of your wax painting with watercolor paints. The wax repels the paint, forcing it into confined areas and giving your painting happy, accidental patterns of color that cannot be predicted. You can give a crayon drawing the same treatment. Artists call this resist painting because the wax resists the watercolor paint.

You can combine wax painting or drawing with the use of dye. Paint or draw on light paper that can be crumpled to fit in a can or bucket of dye solution. Make the solution with all-purpose textile dye, the same kind you used in making lake pigments with chalk. The unwaxed portions of the paper will

Don't hesitate to experiment. Here a watercolor wash over a fish painted with melted crayons gives pleasing results.

soak up the dye and the folded or creased lines will show up as dark webs. The waxed portions will repel the dye.

If you wish, you can remove the wax entirely by placing the painting between newspapers and ironing out the wax. As the iron heats the wax, it melts and is soaked up by the porous newsprint.

You can also scratch through wax paint, showing the white paper beneath to produce a design. Stiff wire such as an unbent paper clip is ideal for this. The scratched painting can then be washed with watercolor or dipped in a dye solution. The results will probably please you.

Art shops carry oil crayons. These too can be used in resist painting. In addition, a drawing with these crayons can be given the appearance of a painting by touching the crayon marks with a brush dipped in turpentine. A heavy application of turpentine will make the colors run. You can, in fact, mix colors right on the paper.

BUYING PAINTS

So far we have been talking about paints you can make. While these lend themselves to some art projects, their permanence cannot be guaranteed. If you are a busy artist and can find just limited time for your hobby, you want to spend that time painting, not grinding and mixing. And of course, if you want to paint furniture, interior walls, or touch up the exterior of your home, you want the right paint for the job. You must go to a paint shop.

The choice of watercolor over oil, or casein over acrylic paints is a matter of individual talent and preference. If you are a beginning artist, however, you will be wise to buy just a few colors at first and test your ability to use them before investing in expensive paint sets and brushes. Also, in nearly all kinds of paint there are at least two grades, student and professional. Start with the cheaper student grade.

Unfortunately, watercolor paints, while the cheapest, are at the same time the hardest to use. You cannot correct mistakes; and brush control, color mixing, and proper handling of absorbent watercolor paper come only with practice. Oils are easier to use, but they are far more expensive. Oils also must be thinned with turpentine and brushes must be cleaned in turpentine or paint thinner. Acrylic paints, not quite as expensive as oils, are water-soluble and extremely versatile, but the beginning artist should not select them as a compromise between watercolor and oil. They have their own distinct qualities which must be discovered and mastered.

The popularity of casein paints has declined recently, and some shops no longer carry them. This is due perhaps to the rise in popularity of acrylics.

Most art paints come in soft metal tubes. When buying them, you can tell by feel if the paint inside is soft. If the tube feels rock-hard, don't buy it. The enemy of all stored paint is air. A tube with a leak or loose cap will eventually harden.

After using paint, replace caps carefully. Sometimes it is necessary to clean out excess paint from the cap threads to be sure of a tight fit. Also, as you use the paint, roll the tube from the bottom, keeping the remaining paint in the tube in a closely packed mass. With proper care, tube-packaged paints can last for years. Store them in a cool spot and above all, do not leave them in the sun. Watercolor paints come in tubes, jars, or sometimes cakes. The tube colors are usually the best quality. Professional artists prefer them also because the tubes provide protection from air. Paints in jars invariably harden with age.

Unless you are very experienced, you will need help when buying a house paint, furniture paint, or paint for any other project in your home or shop. The best source of help is the paint store owner. He has to keep his customers satisfied. He does this by selling sound advice along with his paint.

Always tell the store owner or his clerk what you plan to paint. You must describe the material—that is, wood, cement, metal; its use—that is, exterior deck, kitchen chair, bathroom wall; and the appearance you desire—that is, gloss or nongloss, color and so on. The owner or clerk will probably have many other questions before he recommends the paint. He may want to know if the surface has previously been painted, and if so, with what kind of paint. You may have to sand the surface before repainting. You may have to remove all the old paint.

How do you plan to apply the paint? With a roller? A brush? Or a spray gun? This can make a difference in the paint you buy. If you are uncertain, be sure to ask for help in deciding the application method and in selecting materials.

In addition to listening to advice when buying paint, always read the label on the can before you complete the purchase. This cannot be overemphasized. If the directions on the can disagree with what you have heard, or if they say something you do not understand, more questions are in order.

BUYING FOR COLOR

One of the most difficult jobs in buying paint is color-matching. Suppose you want to paint a new bookshelf to match the walls of a room. You want to use enamel. The walls have been painted with an acrylic paint. Most large paint manufacturers put out standard color charts for several types of paint. One company, for instance, has a tan enamel that matches a tan acrylic, a blue enamel that matches a blue acrylic, and so on. Even so, you must remember that colors fade with age. Don't expect the new paint on your bookshelf to look exactly like the old paint on your walls. In addition, no matter how carefully each batch is mixed, colors often vary slightly from one paint can to the next. This sometimes shows up with distressing contrast when you run out of paint halfway through painting a wall. The fresh can you use to complete the job might be just a shade different, but the difference will show. If possible, plan your job so you make the change at corners, not in the middle of a wall.

The store's color guides with their small circular or rectangular samples of color are, of course, indespensible in choosing colors, but they can be misleading. A color that pleases you in a sample the size of a fifty-cent piece can overpower you when it covers an entire wall. As a general rule, use caution in selecting strong tones. Samples of paint that look pale or washed out in the color guide can be just right when spread over the walls of a room.

PAINTING SURFACE

Artists call the surface on which they paint the "ground." Preparing a ground is of top importance, both in appearance of the final painting and in its durability. Oil painters using canvas must first seal the canvas with glue solutions or an undercoating called gesso, a white paint using one of the extender pigments. Without such preparations, oil would soak into the canvas in such volumes that the paint would deteriorate rapidly. Watercolor paints, on the other hand, require special absorbent papers to assure rapid drying. With many of these papers, a board backing is necessary to prevent curling or buckling during the painting process.

Surface preparation in home painting projects is just as important as it is with art projects. Surfaces must first of all be clean and dry. Dirt will not only smear and stain paint, but it will also cause rapid deterioration of the coating. Moisture is the chief cause of blistering, particularly on wooden surfaces.

Putty and other preparations are available for filling cracks, knotholes, or scars in surfaces. These should be sanded smooth after the filler preparation has dried. In some cases, you will have to sand the entire surface before painting. This is particularly true when painting furniture or cabinets. You must sand between successive coatings of varnishes and enamels to give a toothed foundation for the fresh layer of paint. Otherwise, the layers will peel off like sunburned skin.

PAINTING A ROOM

It would take too much space to list all the common home painting projects, but a description of one will give an idea of what is involved in planning any job, how time can be saved and mistakes avoided, and how you can get the best service from the paint you use.

Suppose you plan to paint a room. Assume you have selected the kind of paint and the color and the method of application you plan to use. The next question is, how much? The store clerk can help you here if you can give the dimensions of the room—the length, width, and height of the ceiling in feet. From the dimensions he can figure the total square footage to be covered and judge very closely the amount of paint you will need. It is best to get a little too much paint rather than too little.

Most shops today have mechanical shakers. Be sure the cans of paint you buy receive a thorough shaking before you leave the store. This will save you much time in hand-stirring.

In preparing the room, the first step is to remove or cover furniture. You can use a drop cloth or newspapers to protect rugs and floors. No matter how careful you are, paint spills and splatters. Next, make sure the walls are clean. In most rooms, light dusting will do the job, but in kitchens it is often necessary to clean grease stains with soap and water. The soap must be rinsed away and the walls given ample time to dry afterward.

Next check areas that are not to receive paint, such as window casings, light fixtures, and floor moldings. Masking tape, sold by the roll in paint stores, will give good protection here. You can paint closely and quickly near taped areas without fear of sloppiness. It is simplest to remove plates around electric light switches and plugs.

When the room is ready, open the paint and use a stick to stir it. All lumps and streaks of color must be stirred out before painting can begin. If the paint has been recently shaken mechanically, the job should be brief, but some paints have to be stirred from time to time during use. Keep the stirring stick handy.

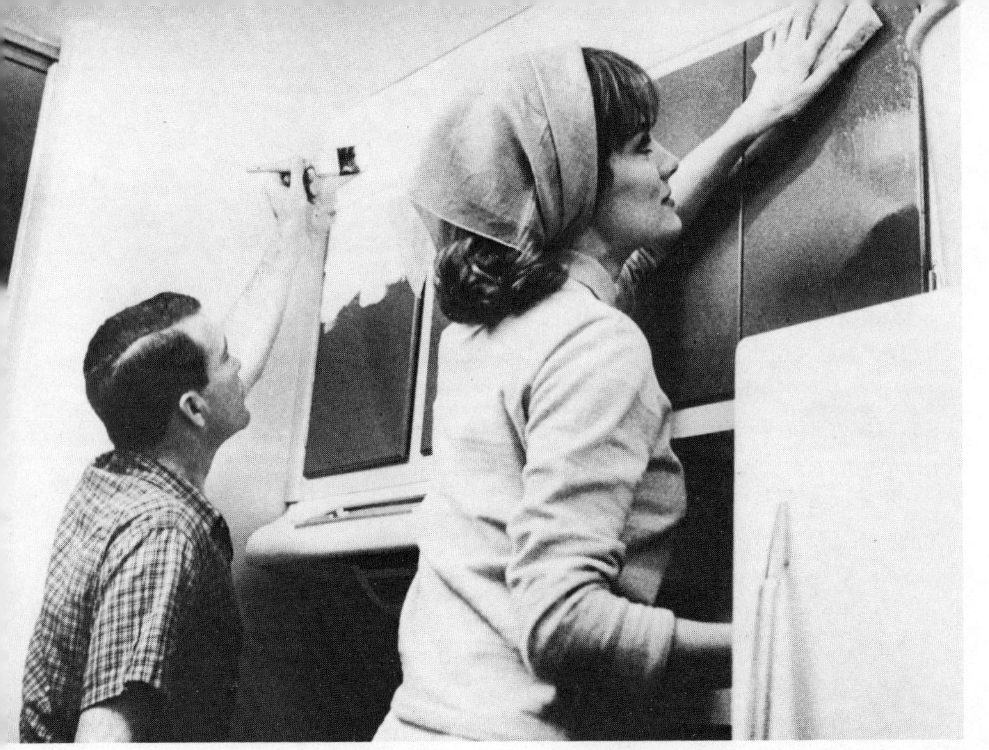

NATIONAL PAINT AND COATINGS ASSN.

Pains must be taken, especially when painting kitchen cabinets, to clean dirt and grease from the surfaces.

Work from top to bottom. If you plan to paint the ceiling, do that first. Paint splattered onto the walls will be covered up as you work down. If you use a brush, dip no more than one third of the bristle length into the paint, and tap the bristles lightly on the inside of the can to remove excess paint. If you use a roller and pan, pour just a little paint at a time into the deep end of the pan. The shallow, sloping end of the pan is designed for spreading the paint to assure even distribution around the roller. With a roller you will still need a brush for narrow areas and corners.

Of course, you will need a stepladder or a stool to reach the ceiling and upper portion of the walls. Be sure whatever you use has a solid foundation and be ready to move it frequently to avoid reaching too far beyond your center of gravity. A fall can bring your painting project to an untimely end.

In painting floor moldings, window casements, and other small areas, you will want to use a small brush. You may also use enamel here. It is recommended for window casements because of its water-resistance. Spread enamel thicker and with a lighter brush stroke than you used with your wall paint.

When painting window sashes, masking tape is extremely useful in bordering window panes to keep paint off the glass. Without tape, you need a *very* steady hand.

Doors are a special problem. Paint the door frames first. Then paint the top and the edges of the door before moving to the broad, flat surfaces. Sometimes it is easier to remove handles and other hardware rather than trying to mask them.

Before cleaning up and putting away your equipment at the end of the job, check your work carefully to make sure no spots have been missed. Use water or recommended thinner to clean brushes, rollers, and roller pans. If there is paint left, seal the lid back firmly. Be sure the groove that receives the lip of the lid is free of paint. Do this by circling it with a dry brush. You can seal the lid most tightly with light hammer taps.

Now gather up the drop cloths, newspapers, and furniture covers. If any drops or specks have stained floors or rugs, wipe them up immediately with a cloth dampened with water or thinner.

At this point, if you inspect your work and find a spot or spots you missed, you will understand why it is important to check before cleaning up. It is a lot of trouble to break out the paint again.

4
Inks — Man's Communicators

Writing developed long before the invention of paper and ink. About 3500 B.C., the Sumerians, who inhabited the fertile plains of Mesopotamia, began to keep records by marking wet clay tablets with a wedge-shaped stick. The hardened tablets bearing this first known writing—cuneiform, it is called—are being translated and studied today more than 5000 years after they were made.

The Sumerians, who also invented the wheel, the plow, the loom, and the potter's wheel, learned to print by rolling a carved cylinder over wet clay. These were clever people, but they did not invent ink. Credit for that is shared by the Chinese and the Egyptians.

The Egyptians had something the Sumerians lacked—paper. It was not like the paper we use today. It was made from the papyrus reed that grew along the banks of the Nile. Strips of the inner pith of this reed were laid out in overlapping sheets and bonded together by pounding. Then, after being allowed to dry, it was polished with a stone to create a smooth and pliant writing surface. This reed paper, also called papyrus (from which we get the word "paper"), first appeared around 3000 B.C. The Egyptian scribes wrote on it with a brush dipped in a mixture of soot, glue, and water—a black ink. They also had a red ink made from ocher, and later they used other pigments, including lapis lazuli. A typical papyrus was about a foot wide and several feet long, designed to be rolled up for storage. The longest found to date measures 135 feet. The Egyptian pictorial writing, called heiroglyphics, can be read today by scholars. Thus the ancient, inked rolls are a remarkable window to history.

The Chinese, like the Sumerians, began writing without pen or paper. Chinese characters, representing single words, phrases, or entire thoughts, have been found on wood or bamboo tablets. They were scratched on the surface with a sharp piece of wood or metal. This method continued until about 250 B.C. It was gradually replaced by lacquer writing. Native lacquer served as the writing fluid, and a pointed stick served as the "pen." Later, lacquer gave way to a mixture of soot, glue, and water little different from the Egyptian ink, and the pointed stick gave way to the brush. Chinese literature credits one Mëng-T'ien with the invention of the bristle brush in 209 B.C. At about the same time, Chinese writers abandoned wood and bamboo tablets in favor of a paper made from silk waste.

Because the inks were so much alike, some historians have speculated upon an early, unrecorded trade between Egyptians and Chinese. It seems more likely, however, that the two civilizations developed inks independently. Certainly the Chinese carried the development further.

FIRST PRINTING

In addition to being written with a stick on wood or bamboo, much of the early Chinese writing, particularly about sacred matters, was carved in stone. Such carvings could produce a print on cloth. The cloth was spread over the stone and rubbed with a lump of charcoal or red ocher. The pigment rubbed off only on those areas backed by the raised characters beneath. The rub-print method continued into the sixth cen-

Even as late as 1929, when this picture first appeared, paper-making in some countries followed the old methods. This scene in Chinese Turkestan shows paper being made from the liquefied bark of willow and mulberry trees. This was poured into calico-covered frames and dried in the sun. This method is essentially the same as that described by Ts'ai-Lun in 105 A.D.

tury A.D. as the chief means of reproducing and distributing sacred writing. It worked on paper as well as cloth.

Ts'ai-Lun, who was in charge of Emperor Ho-Ti's arsenal, wrote about men making paper in 105 A.D. It is the earliest description we have of a method close to the modern paper-making process. Tree bark, hemp, and rags were pounded, soaked, and boiled in water to make a mush. Then the mush was poured over screens of bamboo lattice and allowed to dry. Some histories credit Ts'ai-Lun with the invention of paper. Actually, he was simply describing a practice which may have been going on for some time.

Impression printing replaced rub-printing toward the end of the sixth century. The new process required several steps and it needed a special ink, much stickier than writing fluid. First, a plank of pear or jujube wood, softened by soaking, was covered with a thin paste. Next, a sheet of paper bearing freshly written characters was pressed face down on the pasted surface. The characters were thus transferred in reverse to the wood. When the paper was peeled away, a craftsman with a carving tool began cutting away the uninked portions of the wood, leaving the characters standing in relief, or raised position. When the carving was completed, fresh ink was brushed on the characters and a clean sheet of paper was spread over the plank. The printer rubbed his palms evenly across the paper to ensure a sharp impression. Then the paper was lifted to reveal an exact reproduction of the original writing.

Thousands of books were made from carved wooden printing plates. The first books dealt with sacred subjects, but by the tenth century, histories, poetry, and general literature appeared. As the books multiplied, literacy increased, and even

Infinite patience is needed to carve Chinese characters in wooden relief printing blocks, but the Chinese have used the patience along with considerable skill to print in this way for 1400 years or more. Like most of them, this plate prints two pages.

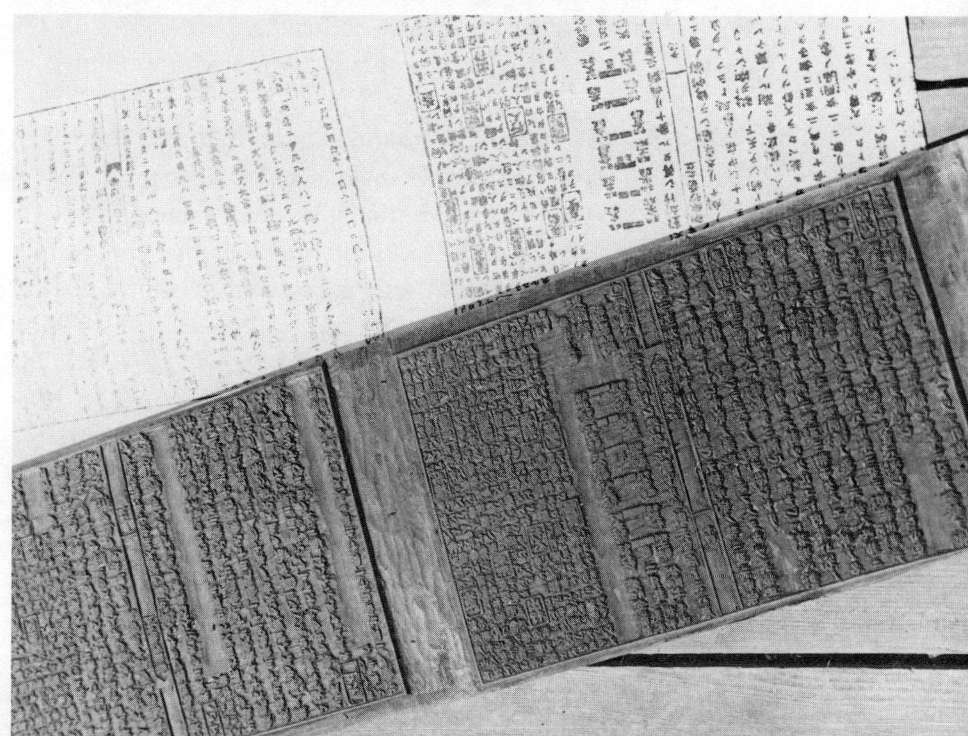

those Chinese who could not read honored and respected the book-makers.

Ink-makers also had great respect. The ink-making art had already advanced greatly by the time printing began. Artists and writers had long demanded the purest black possible. Chao Shwo-chi, writing in 1000 A.D., described how fine ink was made. Oil squeezed from the seed of the tung tree was burned in small terra cotta lamps. The lamps burned within larger chambers which were pierced with small holes that limited the supply of oxygen. This produced a smoky, soot-laden flame. The tops of the chambers had water-filled depressions. The water cooled the chamber to encourage condensation of the soot. Workers lifted the chambers and used feathers to brush up and collect the soot, or lampblack.

Ink-makers ground the lampblack to a fine powder and sifted it through silk screens. Then to each pound of lampblack was added five ounces of glue made either from boiled hides or from stag horn, the whites of five eggs, an ounce of cinnabar, and an ounce of musk, a powerful perfume extracted from certain glands of the musk deer. In addition a small measure of sap from the *ts'in-pi* tree was added to the recipe. Then the real work began. A worker with a blunt stick pounded the mixture again and again in an iron pot. Chao Shwo-chi said it should be pounded at least 30,000 times, but that added poundings would make a much finer ink.

The mixture, which eventually became a jellied mass, was finally removed from the pot and molded into three-ounce cakes, cakes that hardened and were eventually wrapped for sale, often in very elaborate packages.

The artist, writer, or printer using this ink controlled the consistency by controlling the amount of water. The cakes were ground in water on a small stone pallet. By limiting the water, printers could make a pasty ink. By adding more water, artists and writers could make an ink that flowed freely from

Chinese inks lend themselves well to the dry-brush technique, in which the bristles are just barely coated with ink and the style is rapid and sketchy. These beautiful trees were done by the famous Japanese artist Hokusai, who lived from 1760 to 1849.

their brushes. Oriental artists today use ink cakes and the same method to prepare their ink.

CHINA'S INFLUENCE

When Marco Polo returned to Venice in 1295, after seventeen years in China, he soon had the unjust reputation of being the biggest liar in Italy, if not all of Europe. No one could believe his tales of Oriental splendor, his descriptions of such things as asbestos, coal, and paper money. Not until other travelers brought back confirmation, well after Polo's death, did his accounts and his book receive credence. While in China, Polo undoubtedly learned about printing, but he never mentioned it, judging perhaps that no one would have believed that either.

Actually, long before Polo's return the Chinese art of making paper was known in Europe. The art spread, not by intent, but by a series of historical accidents. It happened that Arabs who captured Samarkand in 751 A.D. took two Chinese craftsmen among their captives. These men knew how to make paper, and they taught the art to the Arabs. In the twelfth century, when Arabs entered Spain, they brought the art with them. It spread to France and eventually to the rest of Europe.

A portion of a Buddhist charm, from the British Museum. This was probably printed in the ninth century A.D. *After some eleven centuries the paper is still mostly intact—a great tribute to the ancient Chinese craftsmen.*

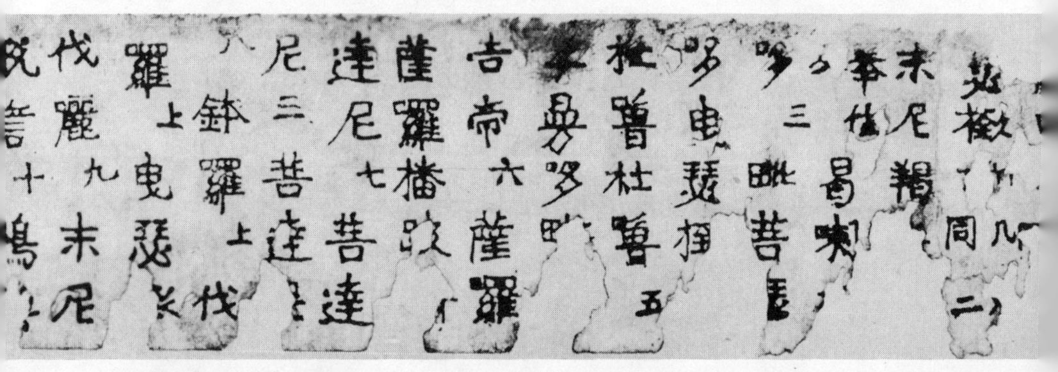

Making paper in 1689. The soggy pulp was evenly distributed on cloth frames, then set aside to dry. This primitive method was still in use in 1929, as can be seen in a preceding picture. The making of inks for printing on such papers was much less critical than ink-making today, when paper is supplied in scores of different weights and finishes, and inks must be adapted to many kinds of presses.

The art of printing, however, was retained in China and its neighboring countries. Europeans had to discover it independently, and long before that discovery writing inks had gone through many stages of development.

OTHER EARLY INKS

The carbon that served as the pigment in the ancient inks made from soot resisted sunlight, moisture, and age so well that many of mankind's early records, as we have seen, survive to this day. The formula for the ink itself survives with few refinements. India inks, as they are known, rely on long-lasting pigments. They are popular among artists and draftsmen.

Unfortunately, the ancients in their search for new colors and different inks often turned to less nearly permanent mixtures. The life of some inks may have been so brief we don't even know about them. These relied on dye for their color. Only a few were successful.

Sepia, the brown-to-black secretion from cuttlefish, was certainly among the earliest dye-base inks. Indigo, another dye, also made a long-lasting ink, but because indigo would dissolve and work as a dye only in certain chemicals, we have to class the indigo used in inks as a pigment. It did not dissolve. The Romans used both sepia and indigo inks and another reddish-purple dye-base that is particularly interesting because of its name. The Romans called it *encaustum,* a word shortened through the years and various cultures to "ink." *Encaustum* came from *enkaiein,* a Greek word meaning "to brand or to burn in." We derived from it the word "encaustic" for hot-wax painting.

Another kind of ink was known in Roman times, but evidently it was rarely used. The naturalist Pliny (23-79 A.D.) wrote that copperas, a compound known to chemists as ferrous sulfate, blackened when mixed with weak acids such as those extracted from nutgalls or barks. The phenomenon led to the development of the gallotannate inks. Though the black was nearly as essentially permanent as the carbon of early inks, the gallotannate inks did not reach wide use until the twelfth century. They had one major fault. The oxidizing process which turned the clear liquid black took from one to two days after it was used on the paper.

Not until some clever but unknown penman began tinting

the liquid with dye did the gallotannate inks become practical. The dye let the writer see what he was writing as he wrote it. Later, by the time the dye faded, the chemical change had occurred to turn the ink black.

Gallotannate inks remained the standard writing fluid through many centuries. American frontiersmen made their own. Abraham Lincoln once recalled how he made good ink by mixing copperas with the juice of blackberry roots.

While inks changed slowly through the centuries, so too did pens. The frayed reed used by Egyptian scribes gave way in Roman times to a hollow reed, pointed and slit at one end, the prototype of the modern pen point. The quill replaced the reed by the Middle Ages. Usually it was a goose feather. The tip was hardened with heat before it was slit and pointed for use. Metal pen points were known in Roman times, but they did not come into wide use until the English manufacturer James Mason began mass production in 1828. A half century later the first fountain pens began to appear.

The evolution of pens and inks proceeded independently up until modern times. The change from sepia to gallotannate, for instance, did not require a change in pens. This was not true in the development of ballpoint pens. The ballpoints, which first appeared in the 1940s, needed special inks. The waxy inks used in the early models failed. They gave the new pens a bad reputation and nearly stopped their development, but early in the 1950s the late Frederick Sayer, a professor at the University of California at Los Angeles, developed a free-flowing ink that stored without drying but did dry almost the instant it touched paper.

Since the introduction of the ballpoint, the fiber-tip and felt-tip pens have appeared. These too require special inks. Most of them are dye-base liquids that use either water or evaporating spirits such as alcohol to dissolve the dye. These new inks are available in many brilliant colors, but for permanence none can match the carbon inks of old.

PRINTING INKS

There are thousands of different printing inks. Press speeds, quality and character of paper, the market for the end product, and a host of other conditions all make different demands on inks; but the basic divisions among printing inks are determined by the printing method.

There are three basic methods. Relief, or letterpress, carries ink to paper from a raised surface. Nonprint areas are depressed so they do not pick up ink and do not transfer it to paper. The Chinese, as we have seen, invented it. Today, more printing is done by relief than by any other method.

In lithographic printing, ink is carried to the paper from a flat surface which has been chemically treated so that part of the surface attracts ink while the other part repels it. Various surfaces can be used, including metal plates, photosensitive gelatin, and stone. It was first done and developed with stone. *Lithography* means "writing on stone."

In gravure printing, ink is carried in lines or dots cut into the printing plate. Before contact with paper, excess ink is wiped from the plate surface so that the only ink left lies in the depressed lines or dots. Under pressure, the paper picks up this ink. This method, also known as etching or engraving, has long been associated with high-quality work.

For convenience in these descriptions, we have spoken only of paper as the carrier of printing. Obviously, many other things from tin cans to road signs can be printed. Also, there are a few other printing methods which either combine or imitate the above or are unique in themselves. They will be discussed later. But first let's look at the three basic methods in some detail.

RELIEF PRINTING

When Chinese craftsmen carved their wooden printing plates, usually putting two book pages on one plate, they were using the technique best suited for their written language. Seeking to save time and labor, Korean and Chinese printers experimented with movable type, but with about 40,000 characters in the language, the experiment proved impractical. A printer would need a warehouse to store his type. Our language, with just 26 letters in the alphabet, could hardly be better suited to a system using movable type, but the first printing in the Western world followed the old, Chinese method. An entire page of text was carved in a single wooden printing plate. European artists who had long been producing woodcut illustrations for hand-lettered books began carving texts on wood blocks early in the fifteenth century, 900 years after the technique was established in China. The European innovation, however, led quickly to the development of movable type. The Dutch and then the German printers began using small wooden cubes, each carved with individual letters. The cubes could be set in line with proper spacing to form words, and the words formed lines of text. The lines, again with spacing, could be wedged into a tray to form the printing plate for a page. Presses with large screws turned by long handles served to transfer ink from the plates to paper. The wooden cubes soon gave way to molded metal type. A new craft of designing and molding type began.

About 1450, Johann Gutenberg of Mainz, Germany, one of those who molded his own type, launched an ambitious project—printing the Bible. Today, the Gutenberg Bible is prized as the first book printed from movable type in the Western Hemisphere. In his day Gutenberg set standards of high quality for the growing craft; he was originally a jeweler,

Johann Gutenberg printed the first book in the Western Hemisphere for which movable type was used. Though printing was not his invention, his introduction of movable type revolutionized what had been an extremely slow, cumbersome process.

with a fine sense of artistry and perfection in details. One of his major contributions was a practical oil-base ink. Because wood type absorbed moisture, it could handle water-base ink reasonably well, but the nonabsorbent metal caused water-base ink to smear under the press. Untreated oils, on the other hand, took hours to dry, but Gutenberg found that boiled oil mixed with small amounts of resin made a workable base. He made his lampblack by burning pitch under iron covers. The ink has held its rich, black color to this day.

If it plese ony man ſpirituel or temporel to bye ony pyes of two and thre comemoracio͞s of ſaliſburi uſe enpryntid after the forme of this preſét lettre whiche ben wel and truly correct, late hym come to weſtmo-neſter in to the almoneſrye at the reed pale and he ſhal haue them good chepe .·.

Supplico ſtet cedula

An advertisement of his services by William Caxton. Try to read the message, bearing in mind that the spelling of the time often looks strange to modern eyes—for example, "enpryntid" for "imprinted," "lettre" for "letter," "westmonester" for "Westminster," and "late hym" for "let him." In the fifteenth century and later, the printed s *somewhat resembled an* f.

In 1474, the first book in English, translated and printed by William Caxton of Westminster, appeared: *The Recuyell* [new compilation] *of the Historyes of Troye.* Caxton, who had learned printing in Germany, used Gutenberg's formula for ink. Unfortunately, other printers were not as careful with their inks. They cheapened them with raw pitch and tallow. Aging turned these inks a muddy brown.

Sometime in the 1530s a printer opened shop in Mexico City, producing the first printed material in the New World. Printing did not begin in colonial America until 1638, when Stephen Daye with his son, Matthew, established the Cambridge Press. Unfortunately, the Dayes and others who followed used poor-quality, imported inks. This practice continued until 1728 when Benjamin Franklin printed his *Pennsylvania Gazette*, using both type and inks he made with his own inventive hands.

Ink-making soon became a special business in the colonies, and by the time of the Revolution there were at least three well established ink firms competing with imported inks. The war cut off the supply of linseed oil, but the ink-makers found

An old engraving of Benjamin Franklin and his press. In those leisurely times the hand-operated press filled his customers' printing needs without the help of today's high-speed motors and rotary printing mechanisms.

a substitute in fish oil. This evidently was adequate, for at the war's end America had 25 growing newspapers.

The early 1800s brought the establishment of several famous United States ink firms. In 1823 a maker by the name of Savage published his technique. Linseed oil, he said, should be cooked over an open fire with constant stirring. After this reached the consistency of thick varnish, it should be allowed to cool and then be mixed with amber or black resin. Then brown resin soap should be added and the mixture once again brought to a boil. After a second cooling, the mixture should be poured into an earthenware vessel containing Prussian blue or indigo. Then, as one worker stirred the mixture, another should add the lampblack. Stirring should continue until a smooth, thick paste formed. By the 1850s makers began experimenting with colored inks, using a variety of pigments other than lampblack. Meanwhile, other changes were being made to improve the printing industry.

The first steam-powered press went into operation in Germany in 1810. An American version appeared in 1827. These were both flat-bed presses with the type locked into page forms and the paper pressed against it or rolled over it. In 1846 Richard March Hoe of New York City invented the rotary press. In this method a cylindrical casting was made of the page and locked upon revolving drums in a high-powered press. Hoe's press was four times faster than the steam-driven, flat-bed presses.

Further improvements made it more efficient and faster. William A. Bullock of Greenville, New York, invented a rotary press that could print on both sides of a sheet of paper, and the paper was fed into the press from a continuous roll. To this design Hoe added a folding machine which cut and folded papers as the press disgorged them. These developments gave the foundation for the modern high-speed presses which produce thousands of copies of magazines or newspapers in a matter of minutes. These presses are so fast they need brakes to prevent them from "coasting" out several hundred excess copies each time they are stopped.

Such speed makes a strict demand upon ink. It must flow freely from storage tank to ink roller to assure even distribution. At the same time, it must dry within fractions of a second upon contact with paper. Wet ink would smear in the folding machines and destroy the work. On absorbent papers such as newsprint the drying process is aided because much moisture goes into the fibers; but on slick papers like those in magazines, special, fast-oxidizing chemicals must be

An old playing card showing a press of the sixteenth century and two ink balls, which were used to apply ink to the type.

mixed with the ink. On some presses hot ink is used. It is of just the right consistency to flow when hot and hardens as soon as it cools in contact with the paper. On other presses, ovens or heat elements bake the ink hard; in some cases the paper must pass—swiftly!—over an open flame. Each method requires a special kind of ink. Obviously, modern ink-making is an exact science.

LITHOGRAPHY

About 1796 Aloys Senefelder of Germany, trying to find a cheap way to reproduce sheet music, came upon lithography by accident. The story goes that he was in his workroom concocting an ink with soap, beeswax, tallow, resin, and lampblack when his mother interrupted him with an annoying request. Had Aloys remembered to take the laundry to the town laundress? No, he had forgotten. She reminded him all over again and told him to list the laundry first to be sure nothing was lost or stolen. Impatiently, he searched for a piece

Aloys Senefelder's first lithographic press, exhibited in the German Museum, Munich. The platen, or flat plate that presses the paper against the stone, is hanging near the right side of the press; foot power applied the necessary pressure.

of paper but could find none. Instead, he used his ink to write the list on a slab of limestone on his workroom table. Some days later, when he again took up his experiments, he tried to clean the ink from the stone. It would not come off. In fact, water would not even stay on the inked portion of the stone. This gave him an idea. He wetted the entire stone with water and then used a roller to spread his ink. Next he pressed paper on the stone and lifted off a reverse print of the laundry list.

It makes a quaint story, but one suspects that Senefelder was more deliberate in his experiments. He certainly lost no time in introducing his discovery to the world. Lithography was particularly suited for posters and handbills. Artists exploited it.

Goya, Delacroix, Daumier, Degas, Manet, and other famous nineteenth-century artists worked extensively with lithography. The Frenchman Henri de Toulouse-Lautrec is best known today for his bold, multicolored posters of Paris night life produced just 100 years after Senefelder accidently produced the reverse image of a laundry list.

Limestone is still considered the best surface for artistic lithography, but commercial printing today is done from zinc or aluminum plates. These metals are chemically treated so that they behave like the stone. The parts that print attract ink while the nonprint portions repel it. Books, magazines, and newspapers can be mass-produced from zinc or aluminum plates on special presses that use offset rollers. The roller with the treated and inked plate does not touch the paper. Instead, it turns against a blank offset roller which picks up the ink, and this in turn presses against the paper. Such an arrangement reduces wear of plates, makes for cleaner impressions and also allows printers to make plates in "straight" copy. That is, the letters, pictures, or designs on the plates do not have to be in reverse; since each roller leaves a reversed impression, the impression on the paper is again the right way around.

In nearly all commercial offset printing the plates are prepared photographically. The common technique involves making a negative of the matter to be printed. Then the metal plate is exposed to light through this negative. The light causes the necessary chemical changes in the plate. Exposed portions attract ink. Unexposed portions repel it.

Lithographic inks must be insoluble in water. Otherwise inks would "bleed" into white areas. Offset presses do not lay ink on as thickly as relief presses; thus the pigments used must be highly concentrated. At the same time, the inks cannot be too sticky. Sticky inks might grab and tear fast-moving paper.

A large lithographic printing plant in Paris during the nineteenth century. There were at least 14 presses in a row down the middle of the room. A "library" of lithographic stones is stored along the walls above and below.

They could even jam a press and do extensive damage, so offset inks must meet strict specifications. Artists working from limestone must be careful to use acid-free inks. The acid would react with the limestone. Even when using commercially prepared lithographic inks, artists often spend hours grinding them to assure full and even concentration of pigment. Some artists prefer to make their inks.

GRAVURE PRINTING

We do not know who invented etching. Scratching lines in wood or stone to create designs was practiced for centuries. Lifting prints from such designs may have first been done on wet clay, perhaps as an accident. Usually, however, when we speak of etching, we are talking about printing from metal plates. Artists began using the method early in the sixteenth century. At first lines were scratched into the metal surface with a sharp tool, a technique we call drypoint today. Later, acid cut the lines.

The metal was first coated with varnish or some other acid-shielding material. Then the artist used a pointed instrument to scratch his design. The scratches removed the varnish. When the artist finished scratching his design, he put the plate in an acid bath. The acid ate into the metal only where the varnish had been removed. The artist could control the depth of the cut through controlling the time he left the metal in the acid. He could make some cuts finer than others by covering them with varnish part way through the cutting process.

After removing the plate from the acid, the artist cleaned off the varnish with a solvent and inked the plate. Before running the plate and a piece of printing paper through his press, he wiped it with an absorbent cloth. This removed the ink from the uncut surface of the plate. Under the press, the paper picked up the ink remaining in the cut portion of the plate.

Prints made by this process have sharp contrast and a wide range of tones, from delicate to bold. The technique has many variations. Aquatint, for instance, uses a soft varnish which allows acid penetration so that the metal beneath it receives a grainy texture. Aquatint prints, which first appeared in 1750, match the soft tones of a watercolor painting.

In 1852 W. H. Fox Talbot of England developed a technique for making printing plates with light-sensitive gelatin as the acid-resistant coating. On exposure under a photographic negative, he could produce line drawings in black and white. Talbot's were relief plates with the portion to be printed raised above the nonprint surface; but in 1878 Karl Klic, a German craftsman, refined the process so that etched plates could be produced photographically. Sixteen years later, Karl Kleitsch of Austria introduced rotogravure, done on a rotary press that used etched plates fitted to cylinders. Today, high-speed rotogravure and similar photogravure presses produce such things as magazines, calendars, greeting cards, and many other materials demanding top-quality reproduction.

On these presses wiper blades constantly scrape excess ink from the rolling plates. A very small piece of grit can scar the plates. Thus great care is needed in producing clean inks for gravure printing. And because the ink is laid on the paper in a very thin layer, the pigments used must have high covering power. In addition, when used on fast presses, the inks must be rapid dryers.

Wedding invitations and calling cards are still printed today from hand-engraved plates. The expensive technique can be recognized by the raised ridges of ink left on the paper, but we shall see that it is possible to be fooled.

PRINTING PHOTOGRAPHS

Talbot's method for making printing plates with light-sensitive gelatin worked well for producing line drawings and other black and white designs, but the shades between black and white, the halftones, remained a problem. Talbot evidently experimented with plate exposures through cloth. The weave of the cloth left a pattern on the plate which seemed gray in a print, but commercial halftone printing did not appear until 1886. The American inventor Frederic E. Ives made plate exposures through special screens that produced dots instead of solid masses. The screens broke the areas of the photograph up into tiny dots, larger dots, and—where the larger dots merged with each other—a crisscross pattern. The light parts of the photograph became small dots on the plate; the gray areas became larger dots; dark or black areas became crisscross. With very fine screens, all dots and crisscross patterns were so small the eye could not detect them without magnification.

There have been many refinements in the process. Louis and Max Levy of Philadelphia made substantial improvements

Magnification shows clearly that a halftone photographic reproduction is broken into dots and crisscross, like this.

in 1893, but the halftone technique which gives us photo reproduction in just about every magazine on the market today has to be credited mostly to Ives. Look at the illustration in a magazine or newspaper with a magnifying glass. This will give you a clear idea of how the technique works.

What about printing colored pictures? Basically, the same dot system is used, but the dots are different colors. Most colored presses use four different inks—blue, red, yellow, black—and thus four different plates. The dots are closely spaced so that the eye does not see them individually, but instead sees them as blended colors. With the white of the paper as a fifth color, the color press can reproduce the rainbow. You can use a magnifying glass again to see how the system works. Sometimes careless printing results in double images in colored pictures. What has happened is that one of the plates is not striking the paper at precisely the right point to register with the printing of the other three plates.

OTHER METHODS

There are many other printing methods. The most important still to be described is stencil printing. Cave artists left both negative and positive hand prints on the walls of their caverns. A positive print was made simply by pressing a paint-coated palm on the stone. The negative print took more imagination. The hand was pressed firmly against the wall while paint was sprayed or squirted around it. When the hand was removed, its shape remained as an unpainted area in painted wall. This is the oldest form of stencil printing.

By placing any kind of pattern on a surface before painting or inking, a stencil print could be produced. It was a simple technique, but strangely the artists of the Orient were long the only ones using it. It took refinements for stencil printing to reach commercial use in the Western world. In

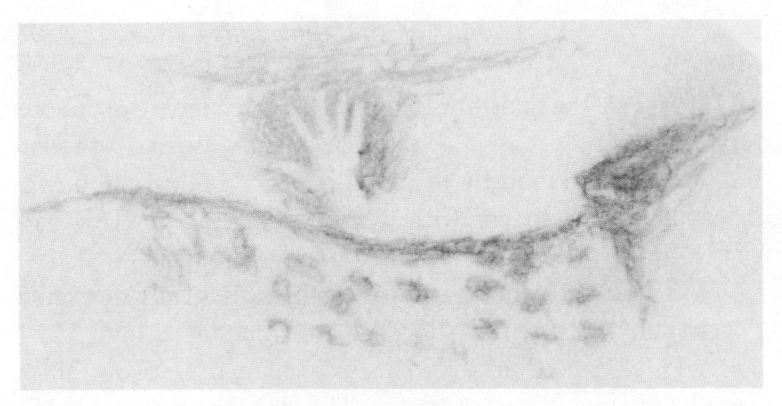

An ancient stencil. This image was painted around the hand of a cave artist in the Peche-Merle prehistoric caves of France. It appears above a horse, only the back of which is shown.

1907 Samuel Simon of England took out patents on a process for printing posters and large signs by forcing ink through a silk screen onto the surface to be printed. By blocking portions of the screen with wax or glue so the ink would not pass, Simon produced stencil prints. Simon's methods were expanded within a few years by John Pilsworth of San Francisco, who made several screens, each for a different color, to produce a multicolored print.

Pilsworth's silk-screen printing was first exploited by industry. Most highway billboards are made by this method. Highway markers and road signs with just two or three bold colors are also best produced by silk-screen stencils. Artists, who usually lead in the development of new techniques, did not take very much interest in silk screen printing until after World War II, though there was a temporary upsurge of interest in the 1930s. Multi-printings of translucent, dye-base inks allow subtle gradations in color that many modern artists use with great success. Commercial silk-screen printers have developed a photo process. Here again light-sensitive gelatin is used. After exposure to light through a negative, those areas to be printed wash from the screen, easily leaving an impervious mask or stencil on the screen for the nonprint areas.

Another widespread application of the stencil printing principle can be found in nearly every major office building in the world. Stencils for mimeograph printing can be cut cheaply with an ordinary typewriter on a special wax-soaked paper. Drawings or graphs can be cut with a sharp tool. After cutting, the stencil is placed on a drum-shaped roller made of wire screen and porous cloth. Ink is forced through the drum and through the cut portions of the stencil to print as the drum is turned.

Hectograph printing, though not as common as mimeograph, provides another cheap method for office reproduction. This is really an extension of the lithograph principle, except that a slab of gelatin serves as the printing plate instead of stone or metal. With a special inked paper as backing, the ordinary typewriter can again be used to prepare a hectograph plate. The ink sticks to the gelatin surface and prints when the plate is turned on a simple press. Hectograph plates give limited printings and the impression is not as sharp as that of mimeograph prints. Most offices use a purple hectograph ink.

Some techniques combine the basic principles. Some offset plates, for instance, are now made in slight relief. By raising the portion to be printed above the rest of the plate, the printer gets sharper impressions.

Heliotype printing, introduced in 1870 by Joseph Albert of Germany, is done with glass plates coated with gelatin. It is an ideal method for mass production of large posters with solid colors and is one of the few printing techniques that does not use the dotted or screened plates for halftones. Most of the posters used to advertise motion pictures are made with heliotype plates.

There is at least one printing method designed to deceive. By using special inks on a relief press and by sprinkling the ink with an absorbent dust before it dries, the printing can produce the raised lines characteristic of expensive, hand-engraved work. It takes an experienced eye to distinguish between an imitation and a real engraving.

By contrast, steel engravings demand top skills of the printers' art and allow absolutely no deceptive techniques. Currency, postage stamps, stocks and bonds, and many other legal documents are printed from steel engraved plates. Steel is so durable that millions of prints can be made without wear. The last impression will look exactly like the first. Steel also allows hard edges and very fine lines, difficult for counterfeiters to match. Currency and most postage stamps are printed with dye-base inks designed to soak into the paper. Such inks last well, and they too discourage counterfeiters. Most governments keep their ink formulas secret.

While steel plates serve their purpose well, there are other printing jobs which require rubber plates. Flexographic printing, as it is called, is especially suited for round objects such as tin cans and for light material such as cellophane packages for food and dry goods. The rubber plates are molded from forms made off metal type. The high-speed, cylindrical presses use a variety of inks, depending on the print surface and desired results. They must be fast-drying, with good covering power.

OTHER INKS

It would take a bigger book than this to describe all the inks in use today. This chapter has attempted only to describe them by class. A few specialized but highly important inks fall outside the ones mentioned so far.

Carbon-paper inks are unique. Nearly all are made by mixing finely ground lampblack or some other source of carbon with hot wax. The wax is spread thinly on one side of light but tough paper. There are a great variety of carbon papers. Some, like those used in a grocery clerk's pad, are designed for one-time use. Others, like those used in a secretary's type-writer, are meant to be used several times. The pigment must be strong on such carbons to stand up to repeated use. Incidentally, the ink can be redistributed on a worn carbon by holding it above a hot electric bulb.

Stamping inks and inks used in typewriter ribbons must have a trait that manufacturers call recovery power. When a type strikes the ribbon, it transfers ink to the paper beneath. The ink must flow from the unused to the used portion of the ribbon to permit repeated use. Stamping inks must flow in the same way to assure even distribution when the rubber stamp is pressed upon the ink pad. Stamp and ribbon inks are usually made with oil-soluble dyes. They too must have strong covering power because the ink coating transferred to the paper is very thin.

5
Ink Projects

You can make writing ink with nails and tea. Soak nongalvanized nails in water overnight or leave them in a hot solution of vinegar for two hours. Drain off the solution and mix it with an equal quantity of strong tea. The ink is ready for use with pen or brush. The ink will be clear at first, then will gradually turn brown as it dries. Chemically speaking, it oxidizes, changing from clear ferrous tannate to brown ferric tannate.

Obviously it is difficult to write or draw with clear ink. A few drops of food coloring or all-purpose liquid textile dye will let you see what you are doing. There are other improvements and variations you can make in this basic ink formula. Instead of soaking nails in vinegar or water, make a solution with copperas, which is also known as ferrous sulfate. Most druggists carry it. You can buy two ounces for about 50 cents. Instead of using tea, try chopping up and soaking oak galls in hot water. Or you can make Lincoln's ink by brewing roots of blackberries to produce the necessary acid solution.

All the above inks, whether made with nails, copperas, tea, oak galls, or roots, are gallotannate inks, no different chemically from inks which served generations of penmen.

You may find these inks too watery for your needs. They

Tea and rust make a workable ink. A few drops of white glue were stirred into this ink to give it a thicker consistency.

can be thickened best with a small amount of egg yolk. Be sure to beat the yolk well before adding it to the ink and when you do add, work gradually. Too much yolk can clog pen points. You can also use a few drops of white glue as a thickener, but this increases the chances of pen clog, particularly if the glue is not thoroughly stirred into the solution before you use the ink.

Making pigment-based inks is challenging because of the fine grinding required. You will have best luck with charcoal willow sticks and egg yolk as a binder. After grinding the sticks thoroughly, add water a little at a time. At first the moist powder will ball up on your mixing stick. Keep adding water until this balling stops. In a separate container break up an egg yolk in a small amount of water. Then add this mixture to the ink. Mix thoroughly and add more water slowly to bring the ink to free-flowing consistency.

Commercial India inks use gum arabic as the binder, but egg ink will give a rich, glossy black, as good as anything you can buy. And like commercial India ink, your home-made ink will leave a permanent stain. Don't spill it. Egg white or white glue can be used instead of yolk, but there is danger again of pen clog. Egg white tends to leave streaky coatings when the ink is brushed on paper.

Except for finer grinding and slightly more water, this ink is no different from the black tempera paint described in Chapter 3. Pigment inks and paints are closely related.

Dye-base inks of many different colors can be made easily with all-purpose liquid textile dyes. Nearly all markets carry these dyes. Mix three parts of liquid dye with one part of well-beaten egg yolk or thoroughly dissolved white glue. By mixing different dyes you can produce offbeat colored inks and subtle tones. One word of caution. These inks dry slowly. Blotting lightens them. It's best to let them dry in their own time.

Invisible inks have long figured in spy stories. There are many chemical solutions which show no color until heated or combined with another solution. Such inks have recently come into use in schools. A student can grade himself on test papers where answers have been printed with an invisible ink. Brushing on a special solution makes the answers appear. You can make a simple invisible ink at home with lemon juice. Write a message with a pen dipped in lemon juice and allow the juice to dry. There will be no visible writing on the paper, but when you warm the paper over a flame or radiator, the message will appear in light brown writing.

MAKING PENS

With a single-edge razor blade or sharp knife, the proper reed or feather, and lots of practice you can produce fine pens for writing or drawing. We must emphasize practice. Penmaking is an art, and you cannot expect perfection on your first attempt. Once mastered, however, the art can bring great satisfaction, and you can produce writing and drawing tools that give distinctive results in today's world of steel and ballpoint pens.

Unfortunately, Europe and the Middle East have the best native reeds, but you can gather usable reeds from river and lake banks and swamps in the United States. Allow them to dry thoroughly. If you do not want to pick your own reeds, ask at a garden supply store for Japanese flower canes. These work just as well, are inexpensive, and need no drying.

Cut the cane into lengths of about eight inches, taking care to leave ends that are free of node growth or imperfections. With tough reeds these and other major cuts are best done with a sharp pocket knife. Lighter trimming can be done with the razor blade. With a stiff wire remove the pith from the inside of the reed. Then at one end make a step cut, starting with a slanting incision about two inches from the tip. Now make a second sloping cut about half an inch from the tip. This is the trickiest cut. It must taper to leave a thin lip at the point of the reed. Use a razor blade to sharpen this lip to a point.

Now you must cut a slit in the point. This is best done by resting the point on a hard piece of cardboard as backing and carefully pressing a razor through the reed. The slit should extend about three-quarters of an inch up the point. Finally, you must trim the point of the pen so that it is even on both sides of the slit. Do this with a razor on the cardboard backing. Naturally, a wide point will give thick lines, a thin point, narrow lines. If you are doing artwork, you will undoubtedly want to cut several reeds of various widths.

Quills are easier to cut. Start with a wing feather from a turkey or a goose. You should be able to obtain an ample supply of feathers from a produce or poultry firm. When feathers come from birds the tips are covered with a thin membrane which must be removed either by scraping or by heat. Heat is best since it hardens the quill, but do not use too much heat. Fill a pan with sand and place it over a low flame. Stick feather points into the sand. Inspect them frequently. When the membrane shrivels, remove the feathers and start a new group.

Unless you are making ornamental quills, cut the feathers to eight-inch lengths and strip the filaments from both sides. Now you are ready to shape the point. Starting one inch from the tip, make a sloping cut that slices off half the barrel. Make a second cut a half inch from the end that leaves a thin point. Now you must slit the point. Start this slit with the blade, cutting about a quarter inch deep. Extend the slit with pressure. This is best done with the tapered handle of a small brush. Hold the pen in your left hand and insert the handle into the pen with the right. Press your left thumbnail an eighth of an inch above the end of the slit. Force the brush handle inward with a slight upward motion. This will extend the slit to the point of thumbnail pressure.

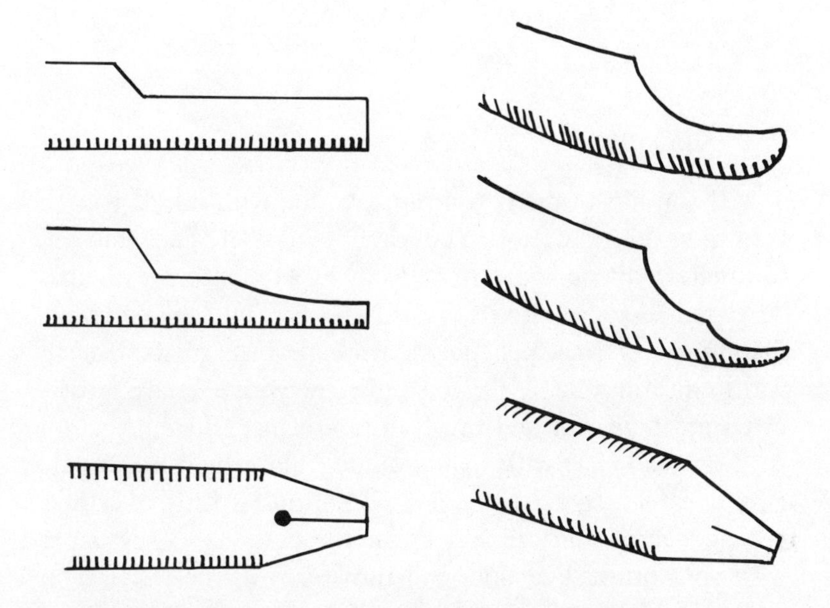

The cuts for a reed pen begin (left, top) *with a slanting cut half-way through the reed about two inches from the tip. The second cut, below it, starts about an inch to an inch and a half from the tip. Bottom picture shows additional side cuts to shape the point. A hole made with a hot pin or a small drill increases the holding capacity of the pen.*

For a quill point (right, top), *the first cut begins about an inch from the tip and the second a half inch from the tip. At bottom, side cuts shape the point. Cutting reed and quill pens takes practice; don't be discouraged if your first attempts fail.*

Trim the sides of the point with a blade and finally cut the tip at a right angle to assure an even point on both sides of the slit. Failures in making reed and quill pens are often due to dull blades. You will be smart to keep a sharpening stone handy and use it frequently. For fine points, a low-powered magnifying glass can be a great help in the finishing work. Sandpaper sometimes serves better than a knife or razor in trimming points.

INK DRAWING

If you like to draw, you can have fun with ink. It is both versatile and restrictive. You cannot erase it, and this unfortunately discourages artists from experimenting with ink. They fail to appreciate its versatile traits. You can use homemade pens or steel pen points available in a great range of widths at an art shop. Or you can draw with a brush. Artists often use both pen and brush in a drawing.

You can start with highly diluted ink, which leaves pale gray lines. The gray makes a good contrast with darker inks you add later. You can wet or puddle portions of the paper with your brush. Pen lines will turn fuzzy in the water. You can use a blotter to control ink patterns on wet paper.

Combining ink with crayons or watercolors gives handsome results. You can work over a watercolor painting with ink after the paint has dried or when it is still wet, depending on the results you want. Try drawing a design with white crayon and then brushing ink over it. The crayon will resist the ink and stand out in sharp contrast. Now, if you wish, you can scratch lines in the crayon and brush on diluted ink. The lines will appear gray. Of course you can use various colored crayons in this technique.

Drawing with a springy stick or twig dipped in ink gives uneven, haphazard lines quite different from the controlled lines of a pen. Some artists draw with chopsticks. Others tie a wad of cloth on the end of a stick, dip the wad in ink, and draw bold lines. Ink invites experimentation.

Felt-tip pens come in many colors and are available in many widths. Most of the colors are pale and translucent on paper and make a charming contrast when combined with black India ink. Fiber-tip pens and ballpoints can also be used with great success in drawing. The Rapido-graph pen,

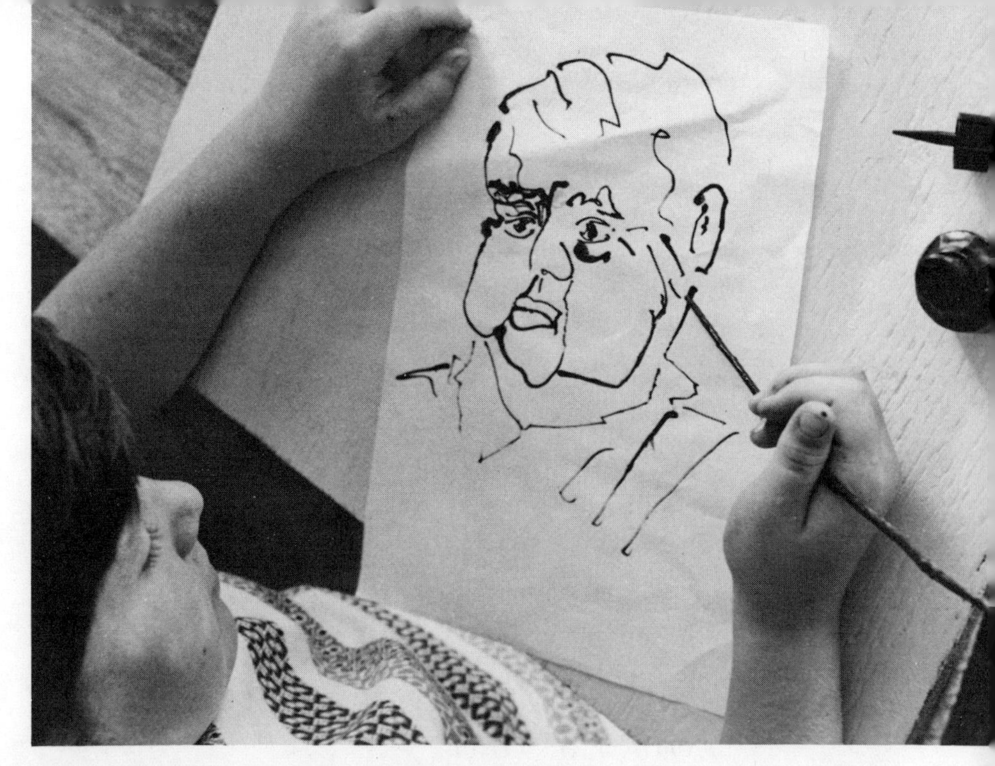

Ink drawing with a stick takes practice, but the end product can be good fun. Notice the varying thickness of the lines.

specially designed for artists and draftsmen, is a fountain pen with a wire tip that gives unique, free-flowing but even lines. The pens are expensive, however, and require a special ink.

PRINTING INK

Making printing inks in enough volume for home printing projects is really not practical. We recommend that you buy tubes of block-printing ink at an art shop. You can get either water-base or oil-base ink. The water-base, though it dries more slowly, is much easier when you clean up after a project. You must use paint thinner in cleaning the materials used with oil-base inks.

In some of the projects listed below, you will be using powdered tempera paints mixed with water. These do not have all the refinements of modern printing inks, but they will serve your needs well.

If you wish, you can make a useful printing ink by grinding charcoal sticks of burnt willow to a fine powder and mixing it in a solution of white glue and water. Make this in approximately the proportion of one part charcoal, one part water, and one part glue. Charcoal absorption varies, however, so you may have to add more water to get a workable, sticky ink. Grind the charcoal thoroughly before adding water. Mix and pour in glue slowly as you continue mixing. You will need a syrupy texture.

It takes a dozen or more charcoal sticks to make a quarter cup of ink, but if you do not mind the trouble, this ink will work well on most of the following projects.

HOME PRINTING

Rub printing, developed by the ancient Chinese, can give great satisfaction to the home craftsman. The texture of rough wood, embossed designs on books, leaf patterns, raised lettering, or designs from any kind of carved surface can be transfered to paper by rub printing. Start with cheap newsprint available in sheets of various sizes at an art shop. Use a black crayon for your first prints. In large work it is best to use masking tape to secure the paper over the work you plan to copy. Then rub the entire paper lightly with the crayon. The letters or design beneath will show through. You can darken your print with increased pressure, but take care, particularly if there are deep indentations in the work below, not to tear the paper. Outside, you can take rub printings from manhole lids, fire hydrant fluting, car license plates, street signs, and letters carved in walls or buildings. One of the best places to go for hand-rubbed prints is the cemetery. Headstones, particularly the old ones, yield a great variety of designs. You may find birds, flowers, and butterflies as well as more conventional angels, stars, and crosses carved in the stones. Use a stiff brush to clean lichen and dirt from the carving before you work.

When you have mastered the technique, you can use more expensive papers. Rice paper is particularly good for rubbings. Also, you can use charcoal or other colored crayons. The rubbings give a fine base for mixed media. A watercolor or ink wash over a crayon rubbing can make a real work of art.

MONOPRINTS

You will need paint or ink, an ink roller, and a piece of glass to make monoprints. As the word suggests, you can produce but one print from your glass plate. To prepare the plate, roll the paint or ink evenly over the glass. Then use a broad-tipped stick to scratch the design in the surface. This will not work if the paint or ink is too runny. Block-printing ink has the ideal consistency. Now press a piece of paper with dimensions bigger than the glass over the plate and rub lightly with the palm of the hand. When you peel the paper off, you will find your design printed on it in reverse.

Why go to this trouble, you may ask, when all you get is one print? The print quality you obtain cannot be matched by any other means. If you are not convinced, try it.

Monoprinting combines well with other techniques such as ink drawing or watercoloring. You can use a watery ink on the plate and produce blotches of thin color, and of course, you can use several different colors on the plate at the same time.

You can make a special ink for monoprinting by mixing powdered tempera paint with evaporated milk. This is actually a casein paint, but it works well as an ink and you can control the consistency easily, particularly if you mix small batches at a time.

RELIEF PRINTING

There are a great many simple ways to produce relief prints at home. The few we list here have been chosen for their ease in obtaining materials as well as variety of results.

You can make a relief plate by gluing pieces of cardboard on another cardboard backing. The cardboard pieces stand up in relief. Ink the plate with a roller. A piece of glass works best for inking. If you use a tube of block print ink, squeeze a little on the glass, then run the roller back and forth until the surface of the roller is thoroughly inked. Then use the roller to apply ink to your plate. Press paper over the inked plate and rub with the hands. Peel the paper off and put it out of the way until it dries. For variety, use the inner, corrugated layer from a cardboard carton. It produces a corduroy texture. For lines, lay glued string on your cardboard plate. Give time for the glue to dry before you print.

Using thick cardboard, you can cut and peel away layers to leave a relief pattern for printing. All cardboard printing has one limiting fault. The ink eventually soaks into the fibers so that by the fifteenth or twentieth print your plate may begin to deteriorate.

Potato printing is ideal for making colorful wrapping paper or wall hangings. Slice a potato in half lengthwise. The inner, smooth surface is your printing plate. Carve away portions you do not want printed to create the design. Cookie cutters are ideal for those who don't like thinking up their own designs. Press the potato's smooth surface well into the cookie cutter and then use a knife to carve away unwanted portions of the potato. Now you can remove the cutter and print.

For ink pads use small kitchen sponges in shallow saucers. Fill the saucers with watery ink made from powdered tempera

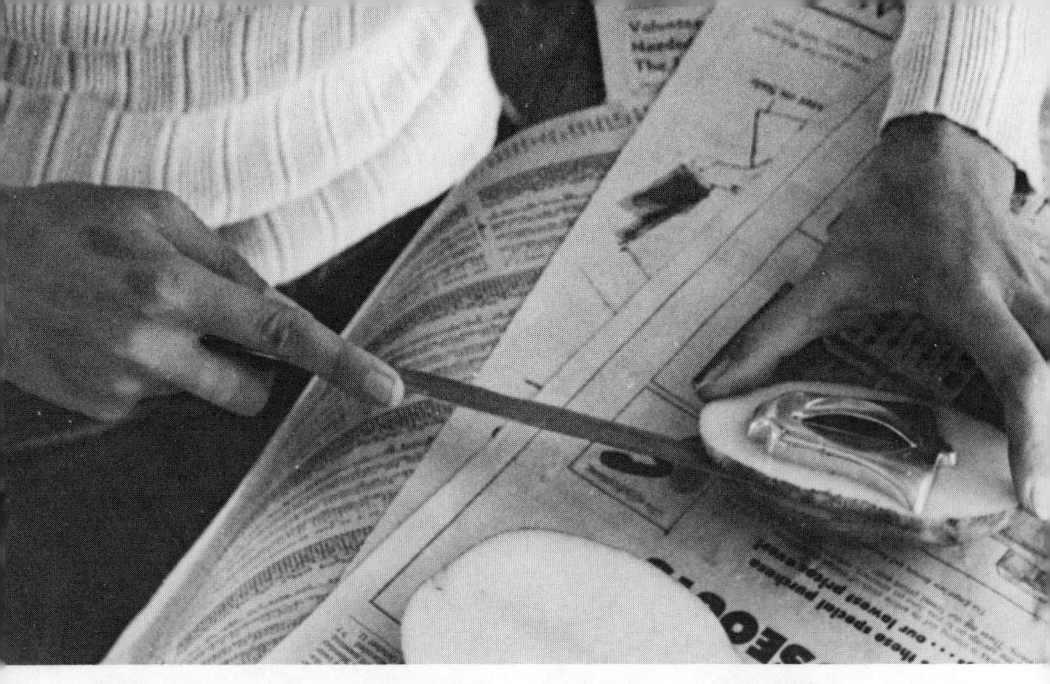

An ordinary cookie cutter, this one shaped like a bell, makes potato printing easy. After pressing the cutter into the flat face of a potato half, carefully cut away a thin layer of potato outside the cutter. Then remove the cutter and you have a printing tool.

The potato printer is best inked with a thick mix of water-base paint, preferably the kind that comes in powdered form. You can control the thickness to suit your needs. A sponge in a shallow pan of paint makes a good stamping pad.

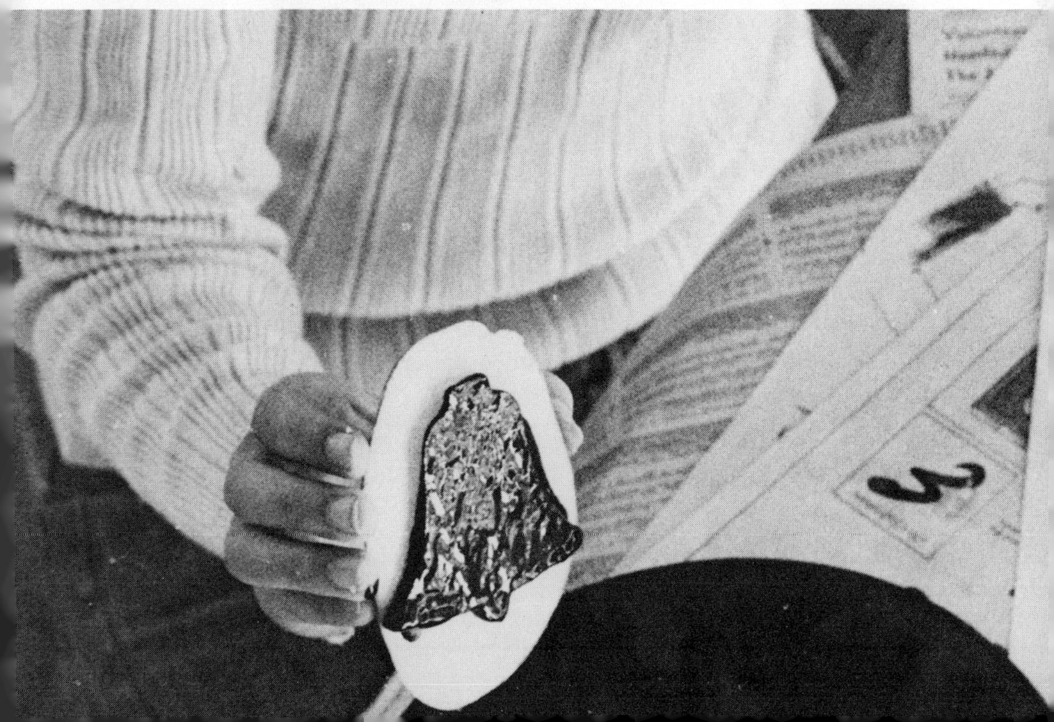

paint. To ink the potato, simply press it firmly on the sponge. Of course, to cover a large area you must use the potato printer repeatedly in lines, circles, or any other pattern of your choice. Bright Christmas wrapping paper can be made with two potatos. One might be cut as a pine tree and the other cut to print a star. You might use green to print the tree and red to print the star in alternate rows across the paper.

The above leads us to sponge printing. Draw your design on a dry sponge, then cut away unwanted portions with a sharp knife. Dip the sponge lightly in a saucer of tempera paint. Then stamp the sponge on a clean piece of paper. As with potato printing, you can pattern the paper with repeated stampings.

You can carve a design in an art gum eraser to make relief prints. This is especially handy for initials or monograms, but remember, when carving letters on the eraser, that they must be in reverse. You will get best results from eraser prints by using block-printing ink and applying it with a small ink roller.

With an old inner tube and a block of wood you can make a relief printing plate far more durable than plates made with cardboard. Use scissors to cut the shapes you wish to print from the inner tube. Detailed cuts can be made with a razor blade or sharp knife. Fasten the pieces onto the wooden block with rubber cement. Make sure they are firm before you ink the plate. Use block-printing ink and apply it with a roller.

With an old rolling pin you can do it a different way, making a "rotary" press. Glue your rubber pieces to the barrel of the rolling pin. You can apply ink with a roller or by running your rolling pin across inked glass. This setup is particularly handy for printing wrapping paper and wall hangings with repeat patterns.

For printing greeting cards or holiday cards, or simply for making art prints, try linoleum block printing. You will need two or three gouges of different widths, a piece of linoleum, an inking glass and roller, a tube of ink and, of course, paper for printing. Art shops stock all these supplies. There is a special linoleum-cutting gouge with several interchangeable points that is very handy. The shops also carry linoleum cut to various sizes and mounted on plywood blocks. There are various grades of hardness. Very hard linoleum gives sharp edges to a print, but it is difficult to cut. We recommend starting with softer grades.

If your linoleum has not been coated with white paint, you can do this yourself. A tempera paint is best for this. Do not use oil paint. The white coating helps in drawing and carving your design. Pencil marks against the natural brown of linoleum are very difficult to see. Remember, if you use lettering in your design, *it must be carved in reverse*; and reversal should be kept in mind as you compose your drawing.

Note: Extreme care must be taken to avoid cutting yourself. The gouges must be sharp. Unfortunately the gouges will slip easily across the slick surface of the linoleum. If you have an old workbench, nail down wooden strips to brace your work. If you cannot do this, always hold the block with your free hand *behind* the gouge, never in front.

In printing from linoleum blocks it is best to have the block face up and lay the paper on it. Rub firmly with the hands or the back of a wooden salad spoon; or use a clean rubber roller if you prefer. There are special hand presses for linoleum blocks, but the expense is hardly justified unless you are printing in great quantity.

Wood is harder to cut than linoleum. You will need a set of wood carver's gouges to do the work properly, but the results with wood can be very pleasing, especially when your prints pick up the natural grain.

You can make a printing plate with plaster of Paris. Mix the plaster in the proportion of three parts of plaster to two

Linoleum makes an ideal printing surface for greeting cards, but you will need special cutting gouges. Here the artist uses a small gouge to trim around letters in her sunburst Christmas card. To prevent an exasperating mistake, remember that the lettering must be backward in order to print normally.

Inking the block is the most important step in getting good results from linoleum printing. Too much ink will gum up details of a design and take very long to dry. Too little ink gives pale, washed-out prints. Good inking takes practice.

parts of water. Make a mold by greasing the inside of a small cardboard box or box top with petroleum jelly. Pour the plaster into the mold and allow at least an hour for it to dry before stripping away the cardboard box. Scratch your design into the plaster (on the smooth bottom surface) with a nail or a knife. When the design is finished, coat the plaster with varnish. When the varnish has dried, you can ink and print. If you plan to make many prints from plaster of Paris, use an oil-base block-printing ink. The water-base inks have a tendency to soak into the block, causing it to soften or crumble.

SMOKE PRINTS

Heavily veined leaves lend themselves best to smoke printing, but you might get excellent results with blades of grass, flowers, or small butterflies. You will need two cylindrical bottles, some petroleum jelly, a candle, old newspapers, and printing paper. Fill one bottle with water, cork or cap it, and smear the sides with petroleum jelly. Now hold the bottle over a candle flame, turning it slowly to collect an even coating of soot.

Now spread the leaf, vein side up, on some newspaper and roll the bottle over it, inking it with the soot. Carefully remove the leaf and lay it on a fresh sheet of newspaper. Put your printing paper over it, and then, with the other bottle, roll the back of the paper to pick up the ink from the leaf. Smoke prints have fine, delicate lines and tones. You should protect them with a spray coating of art varnish, applied from sufficient distance so the spray blast doesn't blow away the soot.

You can also print from leaves by using an inked roller in place of the soot-coated bottle, but you can come closer to the delicate tones of a smoke print by inking the leaf with carbon paper. Smear the veined side of the leaf with lard or

The first step in making a smoke print is "inking" the bottle. A thin film of petroleum jelly is smeared on the glass. Then soot is gathered from the candle flame. The petroleum jelly is the vehicle, the soot is the pigment, in this ink. Keep the bottle moving. A little water inside it guards against cracking.

Leaves of the Japanese maple make delicate traces on the paper in a smoke print. If you want to save your prints, it's a good idea to spray them with clear art varnish. A light mist of varnish will prevent smudging.

petroleum jelly. Place it on newspaper vein side up and put a sheet of carbon paper over it, with another sheet of plain paper as backing. Rub the paper carefully to transfer carbon to the leaf. Too much pressure will crush the leaf. Now place the leaf on a clean piece of newspaper and print as you did before.

STENCIL PRINTS

Cut a design in stiff paper, lay it on a sheet of plain paper and daub on ink. When you lift the stiff paper, you will see the outline of your design. Stencil printing is as simple as that, but it allows many variations. Instead of daubing the ink with a stiff brush, you can spatter it on by using an old toothbrush. Put ink or paint on the brush, hold it above the stencil-masked paper with the bristles parallel to the paper and run the edge of a knife upward along the bristles. India ink works best for this, but you can get good results with tempera paints, and they are much easier to clean up. Try white ink or paint on colored papers for interesting variety.

Leaves, flowers, and other things from nature can be used in place of a cut stencil. Leaves often must be pinned or weighted down to assure full contact with the paper. Here is an ideal opportunity for mixing techniques. When making an ink or smoke print with a leaf, try splattering a contrasting color on the paper before you lift the leaf. This will combine relief and stencil printing.

Sponges, small wads of cloth, or crumpled paper make ideal daubers in stenciling. Each gives a different texture to the finished print. Using different daubers on the same print heightens interest.

Pressurized cans of spray paint can be used very effectively with stencils. Light touches of different colors with changes in pattern with each change in color give excellent results. The

Spattering ink over leaves is a simple form of stencil printing. The bristles of an inked toothbrush over a knife edge give plenty of spatter; holding the brush vertically is usually best. Some artists like the large blotches that drip from the brush; some don't. If you want to avoid them, dip up less ink or get rid of them over scrap paper.

metallic sprays such as gold, silver, and copper are well suited for this technique.

As with all home printing, much of the fun with stencils comes from experimenting. If you think of something new, try it.

6

Dyes—from Snails to Synthetics

When dyes are mentioned, we think at once of cloth, but dyes color many other things too, including leather, paper, human hair, and animal fur. Dyes provide science with an important tool, particularly in work with microscopes. There are dyes which stain certain tissues and leave others uncolored, an important technique in making microscope slides.

Today nearly all dyes are made synthetically. The raw materials are selected chemicals, particularly those containing carbon. These chemicals, combined under carefully controlled processes, become the working dyes of industry. This situation is new, barely a century old.

It used to be that nearly all dyes came from plants and animals. There were no other sources. Farmers harvested leaves, flowers, roots, or bark from various plants that they cultivated on vast acreages just to get dye. In some regions workers patiently collected tiny insects for dye. In remote forests axemen hewed trees into logs. Some of these logs traveled halfway around the globe to supply textile centers with dye.

The labor, shipment, and storage of these bulky dyestuffs made them expensive, but that was not their only fault. Different climate and different harvesting times and methods made the natural dyes inconsistent. No dyer could guarantee the same color from one dye batch to the next. Impurities dulled colors, and sometimes the dyes faded from the cloth with disappointing speed.

Despite the expense, despite the faults, natural dyes practically shaped the commerce of the world. Since the first camel caravans traveled the desert sands, since the first merchant ships sailed for distant shores, dyestuffs, along with silks, spices, and other rarities, dominated the cargoes.

The demand for color, as we have seen, was no less in prehistoric times than it is today. The Chinese began using dyes about 3000 B.C. Egyptians took up the craft soon after. In Europe the Swiss Lake Dwellers who built their homes on stilts above the water to avoid perils of the land practiced dyeing as early as 2000 B.C.

In Biblical times and probably earlier, before the written record began, huge caravans with thousands of camels carried dyestuff from Samarkand and Babylon across the desert to rich markets in Jerusalem and Cairo. Homing pigeons, released daily, carried news of the caravans' progress to waiting merchants.

The search for one dye—royal purple—sparked the exploration and subsequent colonization of much of the Mediterranean world. Phoenicians, bold sailors and crafty merchants, spread from their capital of Tyre on the Mediterranean's eastern shore in search of a marine mollusk, the Murex snail, that yielded the rare dye. The Phoenicians' enterprise led not only to a monopoly in the dye but also to the dominance of all trade, a dominance they held from 1500 B.C. until the rise of Rome. The Phoenicians long challenged Rome and struggled against her power until the fall of Tyre in 638 A.D.

Turkey once controlled the supply of red cloth made with dye extracted from the madder plant. Other regions could grow madder; it has been found on cloth in Egyptian mummy cases. But the Turks mastered its use and kept their methods to themselves as a profitable trade secret. Today the same color

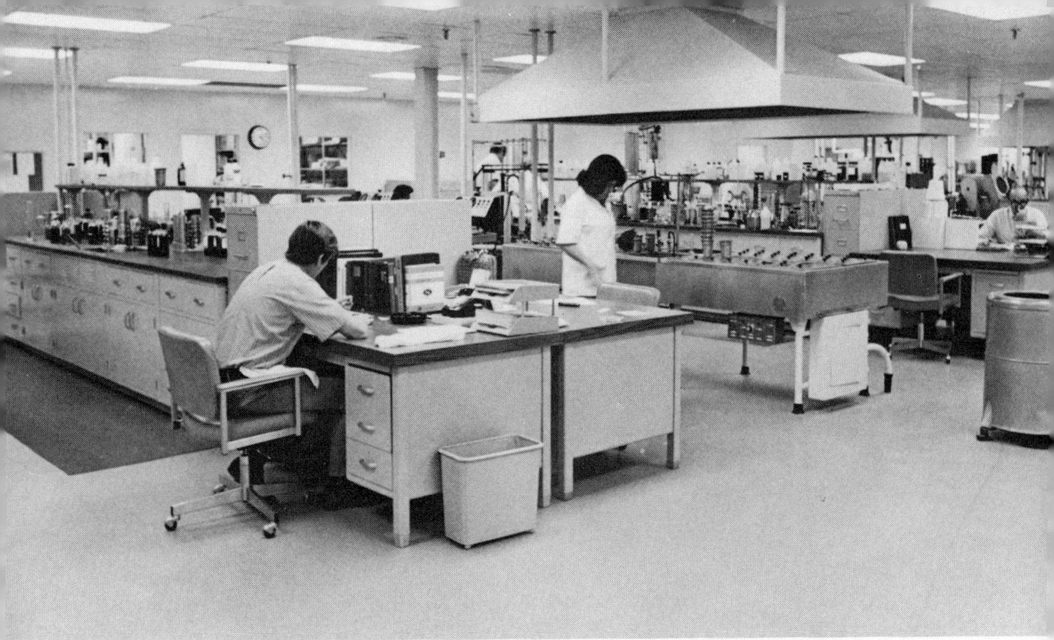

ALLIED CHEMICAL CORP.

New dyes are developed and tested in spotless laboratories, far different from the cluttered dye shops of yesterday.

is produced by other means without the aid of the madder plant or secret processes.

In fact all the natural dyes, as important and valuable as they were, today have practically no value. What happened?

A revolutionary change struck the dye industry. It started in 1856 when an 18-year-old chemistry student working in his home laboratory near London, England, accidentally made a violet dye from an extract of coal tar. It was the first synthetic dye. Thousands of others followed.

Chemists today can manipulate complex carbon compounds to make dyes of any color, dyes without impurities, dyes that do not vary with climate, or growing and processing methods—most of them cheaper than the natural dyes. It is little wonder that by the end of World War I, synthetics had driven natural dyes from the market.

Though they lost their commercial value, natural dyes remain highly interesting to historians, and more and more home craftsmen are using them with enthusiasm. The home craftsmen like them for the very reason industry abandoned them: they cannot be standardized. Each batch makes a unique color and provides a chance to be different.

We will discuss natural dyes, grouping them by color, but first we must look at an early development, one as revolutionary in its day as synthetics are to our era. It happened over 4000 years ago.

MORDANTS

At first men and women protected themselves against the cold with animal skins. The stiff, scratchy skins became comfortable only after long usage, about the time they were beginning to wear out. It must have been a great boon to body comfort when the ancients learned to make leather by soaking hides in solutions of vegetable acids. Bark or leaves yielding tannic acid worked best. Tanning leather, of course, developed long before the ancients understood the chemistry involved,

A logwood-cutting machine from an old-time dye shop; it reduced blocks of logwood to usable chips. This and four other engravings in this series were originally published in Tomlinson's Cyclopaedia of Useful Arts, *London, 1854.*

SMITHSONIAN INSTITUTION PRESS

but they were good observers. They chose the materials that worked best and they stuck with them. And it did not take careful observation when dyeing began to learn that leather held color much better than wool. History does not record who made the important discovery or when it was made, but the craftsman who reasoned that wool might hold dye better if it too was soaked in tanning solution became the first to use a mordant.

The word comes from *mordere*, the Latin word for "bite." A mordant does aid the dye in biting the cloth, but in terms of modern chemistry it is more proper to say that the mordant provides a bond between the cloth and the dye. It is a chemical that reacts with both cloth and dye and thus bonds them firmly together.

The first historical mention of mordants comes from the Middle Kingdom of Egypt, an era that began in 2000 B.C. The Egyptian dyers used alum, a natural salt containing potassium, aluminum, sulfur, and oxygen. It is one of several metallic salts still used by home dyers today. Salts of chrome, tin, copper, iron, and even common table salt, sodium chloride, serve as mordants. Vinegar, lime, caustic soda, ammonia, cream of tartar, and tannic acid also work. The choice of mordants is broad, and it is as important as the choice of the dye itself.

The early dyers found that some mordants worked better in one cloth than another; some were ideal for one dye but ineffective with another, and the dyes sometimes gave different colors with different mordants. Often much depended on how one used the mordant.

It took more than madder to produce "Turkey red." The cloth was soaked in solutions of oil, oak galls, alum, cow dung, sheep's blood, and intestines before it went into the dye bath. Even after the secrets of the process leaked out, the Turkish dyers retained a virtual monopoly on the color. European dyers considered the whole thing too much trouble.

SMITHSONIAN INSTITUTION PRESS

Logwood sawdust is soaked before being put into dye vats.

In medieval times merchants in Genoa, Italy, managed to gain control of all known sources of alum. The monopoly brought them wealth, but it also prompted experimentation among dyers. Tradition called for alum, but when dyers found that other salts could be substituted, the mordant monopoly crumbled.

As we take up discussion of individual dyes, we shall mention mordants again and again. They were extremely important. However, there were a few dyes that worked without them.

BLUES

Indigo was one of these. It was extracted from a plant of the same name that grew wild in India and Egypt. Of course farmers began cultivating it as soon as it gained economic value. Thus it was not only one of the world's first dyes but also one of the first cash crops. The preparation and use of indigo required several steps, but the work was worth it. Dyers could produce many different shades of blue, and the color held fast in cloth without the aid of a mordant. In a way indigo acted as its own mordant.

To prepare indigo leafy stems of the plant, a relative of the pea, were soaked for several days in water, long enough for the solution to ferment. This was an oxidation process. When it was complete, remains of the plants were removed and the solution was stirred for several hours. When stirring ceased, a blue solid settled to the bottom of the tank. Eventually the liquid was drawn off, the solid collected, shaped into cakes, and dried. Properly stored, the caked indigo would last for years. This was how it was shipped from growing regions to textile centers.

The caked indigo would not dissolve in water unless it was treated with special chemicals known as reducing agents. Soap, caustic soda, and other alkaline solutions worked as reducing agents. A fermenting solution could also dissolve the dye. While the methods differed, the results were the same. The blue dye, once reduced, turned pale yellow in solution. It was strange that a yellow solution could produce blue color, but this was indigo's unusual property. Dyers dipped cloth into the solution and then hung the cloth to dry. In contact with air, the indigo oxidized and turned blue, a blue that would not dissolve. Repeated dippings in solution produced deeper shades of blue.

Dyers considered the indigo of India the best. Competition for its control led in part to confrontations between Portuguese, Dutch, and English traders in the sixteenth century.

In colonial America farmers planted both a native indigo and indigo imported from India and Egypt. The crops grew

SMITHSONIAN INSTITUTION PRESS

The earliest known picture of an American dye operation shows a worker at left lifting a dye bucket from a hot vat while others carry dye, bend over vats, and hang finished cloth to dry. This woodcut appeared in Hazen's The Panorama of the Professions and Trades, *Philadelphia, 1836.*

well, but the farmers could not match the harvesting and processing techniques of India. Tobacco and cotton soon became the leading cash crops in America.

Chemists did not succeed in making indigo synthetically until the 1870s, and large-scale production of synthetic indigo did not get under way until the beginning of this century. The compound that gives the plant and the synthetic the blue color is known today as indigotin. By manipulating its composition, chemists can make indigotin dyes of several different colors.

Another early native blue came from woad, a yellow-flowered member of the mustard family. Woad too contained indigotin. The plant grew wild throughout Europe, North Africa, and Asia. Egyptians were probably the first to cultivate it. In 55 B.C., when Julius Caesar invaded England, he faced blue defenders. The native warriors covered their bodies with the juice of the woad plant.

Using woad in cloth was another matter. The leaves were first ground to a pulp. The pulp was then rolled into balls about the size of a soft baseball and dried. Proper drying usually took about four weeks. Then the balls were ground to powder, and the powder was spread on a stone floor and dampened with water. For nine weeks the soggy mixture fermented, turning gradually to a clay-like paste about one-ninth the bulk of the original harvest. This was the dye. Unlike indigo, woad needed a mordant, usually alum. It was mixed in solution with the dye, and the solution was heated for at least three hours before the cloth went into it. The color appeared as the cloth dried.

The deep blues of indigo could not be achieved with woad, but it was nonetheless an important natural dye. Some dyers used a mixture of woad and indigo, partly to cheapen their dye and partly because woad started the fermentation necessary to reduce the indigo.

Prussian blue, one of the first manufactured dyes, appeared early in the eighteenth century. It was not a wild success. It was made from an inorganic potassium compound, the same as that used in paint pigment, and required a mordant of iron salt to work as a dye. The color darkened badly in wool,

SMITHSONIAN INSTITUTION PRESS

An old-time mordanting operation. At left center the unmordanted cloth is drawn into an alum vat on the left, rolled overhead, drawn through a wringer, then stacked in folds on the right.

and while Prussian blue resisted light well in cotton and silk, it washed away in soap.

REDS

Madder, related to the coffee plant of South America, is a native perennial of Asia Minor. Its roots were used to make a red dye from prehistoric times. Cultivation of madder first appeared in India and Egypt. Farmers waited until the plants were three years old before harvesting them. The French were the first to take up madder cultivation in Europe, but the Dutch adopted it and soon excelled in production.

Dye production under the Dutch method called for drying the raw roots in ovens as the first step. Then the roots were pounded repeatedly. The outer husks were removed to be sold for cheap dye extracts. Then the layer beneath the husk was peeled away for another extract not quite so cheap. The inner portion of the root was retained for quality red dye. It was crushed to a powder, packed in wooden casks for aging, and held from one to two years before shipment to textile markets throughout the world. There were crooks in the madder trade. Buyers had to take care that the casks of dye they bought were not corrupted with ground almond shells, brick dust, or red sand.

On America's eastern seaboard the climate favored madder cultivation, but local production never got started, despite the urgings of Dolly Madison and Thomas Jefferson. Dyers were content to import madder, and there were some dyers who resisted madder entirely, saying its use was too much trouble. One dyer said it would take the facilities of his entire shop and sixteen working days to match the Turkey red of Asia Minor.

In the eighteenth century madder became one of the first dyes of medical research when Henri-Louis Duhamel (1700-1781), a chemist and botanist, fed madder to growing pigs. He found that cartilage remained white, but as soon as it turned to bone it turned red, taking on the color of the dye. With young pigs the color was deepest in the middle of bones. With adult pigs the color was deepest on the bone surface. By coloring new bone best, the madder gave scientists their first insight into bone growth and bone repair.

We know today that the red of madder is alizarin. Like

indigotin it now constitutes a special group among the synthetics, with many different shades of color.

Another early red came from lice, not body lice but shield lice that thrived on the holm and kermes oaks of the Mediterranean world. Only the female of the species yielded dye. It required a great deal of labor and patience to harvest the tiny animals, but kermes, as the dye was called, was mentioned in a record written in 1727 B.C. It worked well in wool, silk, and leather. In Greece a traditional red cap, and in the Islamic world the red fez were both dyed with kermes.

The interior of an eighteenth-century French dye shop, with silk yarn being processed. This woodcut first appeared in the Encyclopédie, ou Dictionnaire Raisonné des Sciences, des Arts et des Métiers, *Paris, 1772.*

SMITHSONIAN INSTITUTION PRESS

The pre-Columbian civilizations of Central America harvested a similar insect from the Nopalea cactus to produce a bright red dye. When Hernán Cortés and his men invaded Mexico in 1518, they recognized the value of the dye at once and launched one of the New World's first exports. Cochineal, as the dye is called, also came from the female insect. Fortunately for the harvesters, the females had no wings, while the winged males flew before the pickers' hands. The new dye found an eager market in Europe, but it was expensive. Seventy thousand insects had to be collected, dried, and crushed to make one pound of dye. From one acre natives could produce 250 pounds of dye for their Spanish masters. In addition moisture could destroy the dye. Great care was required in storing and shipping, and buyers had to guard against dye weighted with rocks and pebbles.

Cochineal continued in use well into the synthetic era. Until the development of azo-scarlet dyes its color could not be matched. The insects are still harvested in Central America for craftsmen and native dyers, but cochineal today has little commercial value.

There were many other native red dyes. Many came from the barks of various trees. Brazilwood had the widest use. Brazil received its name because of the vast stands of dye-yielding trees found there by explorers in 1500. The forests were harvested and the logs shipped to Europe by the boatload. Sappan, which grew in India, Ceylon, and Malaya, was also harvested for its red dye. In addition there was camwood, barwood, and sanders wood, all collected with great labor from remote regions to supply the dye vats of Europe. Turning logs into dye required heavy labor. First the logs were chipped, and then the chips were soaked for days on end. Great care was needed in filtering the solution, but often the finished cloth contained annoying splinters.

Safflower, a type native to Egypt, provided a pink dye far

cheaper than others because the thistle-like heads of the plants had only to be soaked in a weak alkali solution to extract the dye. Safflower was used extensively throughout Europe for centuries and served in more than one city to dye the ribbon that bound official documents. The practice gave us the term "red tape" to describe the complexities of law or government.

PURPLE

The Greeks called the men from Tyre Phoenicians from the word *phoenicia*, meaning "red." From this and early descriptions we know that the dye extracted from certain snail shells was a reddish purple. Some accounts call it carmine or magenta, but for most of the ancient world it was royal purple. Why royal? It took 336,000 shells to produce one ounce of dye. Who but kings could afford such a color?

The Phoenicians were remarkable traders and sailors, but they did not go in for art of literature. Consequently we know much less about them than we do about the ancient Egyptians or Greeks. We do know that by 1250 B.C. they had established colonies in Italy, Spain, and North Africa. Wherever they found the dye-yielding snails they landed a crew of men to set up dye shops. This was how their great city of Carthage on the North African coast began.

There were actually two species of snails that produced dye—*Murex brandaris* and *Murex trunculus*. Each had a small gland that secreted dye. A dye bath made from the secretion was pale green, but after cloth was soaked in it and hung to dry, the cloth turned dull purple. Then after soaking in a weak solution of soap, the cloth brightened to magenta.* The dye

* Though the word "purple" almost always means violet today, it is useful to know that it once referred to a magenta-like red (royal purple) rather than a reddish blue (violet).

thus behaved much like indigo, and we know today that the basic compound involved was the same. Impurities caused the differences in color.

Competing merchants tried to match Phoenician dye with a dye made from crushed lichens or one made by mixing red and blue dyes. The attempts had little success. Unscrupulous traders thinned the dye with honey or lichen extract and promptly lost their reputations. Nothing could match the pure dye. Today the Phoenician royal purple, like other natural dyes, can be matched by synthetics.

The marine lichen used to dilute royal purple gave a dye called orchil. The lichen itself, called Roccella, grew on the Mediterranean coast and provided the ancients with a cheap but dull substitute purple. Roccella has also been found on the Cape Verde Islands, the Canaries, and on the coasts of India and Ceylon. The lichen had to be soaked for a week or more in an alkaline solution to extract the dye. By varying the soaking period or the strength of the solution, different shades ranging from blue to crimson-purple could be obtained. Orchil, which settled to the bottom of the vat, was dried and caked for shipment. Used with alum as the mordant, orchil produced red. With an iron salt it turned cloth reddish-purple.

Another purple made from several lichens native to the British Islands was called cudbear. It was patented in Scotland in 1758 and shipped as a dry powder to world markets for many years as a valuable export.

YELLOWS

Weld, an annual plant of Europe and Asia Minor, ranks with indigo and madder as one of the early vegetable dyes. When properly mordanted, cloth dyed in weld turned bright yellow. Most of the dye was concentrated in the leaves, but usually the entire plant was soaked in solution to extract the dye. Several hundred plants were needed for one piece of cloth. This was weld's major fault. It was bulky to store and ship, but dyers demanded it just the same. Weld was very versatile, with different mordants and different textiles producing various colors. Chrome salt, for instance, turned weld olive-yellow in wool and cotton. Alum gave canary yellow. Iron produced tan.

Saffron, a member of the iris family native to Asia Minor, may have been used as a dye before weld. The dye came from the central portion of the saffron flower. It took nearly 4000 flowers to make one ounce of dye. It was costly but it was also very popular, particularly in India, where Buddhist monks adopted the color for their robes. Saffron yellow remains as the traditional mark of priesthood in India today.

Another yellow came from a species of mulberry found in Mexico, Cuba, and Nicaragua. Logs of fustic, as it was called, provided another early export from the New World. Fustic yielded an orange-yellow when alum and cream of tartar were used as the mordant. Dyers chipped the logs and tied the chips in bags to prevent damage to the cloth. Chips were soaked for at least three days. Weak mordant solutions led to early fading with fustic, but if the mordant was too strong, the color would brown badly with aging.

Still another yellow dye, quercitron from the bark of the North American black oak, provided another important New World export. The pioneers also obtained yellows from two members of the buckwheat family—dock and a plant known as arsesmart (which was prickly). The bark of the white ash and the root of the barberry bush also provided important sources of yellow in pioneer America. Only quercitron, however, became an important yellow dye export.

Dyer's broom, which grew well in England, might have become an important crop in America, but our dyers resisted the local broom started here in favor of imported dye.

Chrome yellow, like Prussian blue, was a mineral dye, one

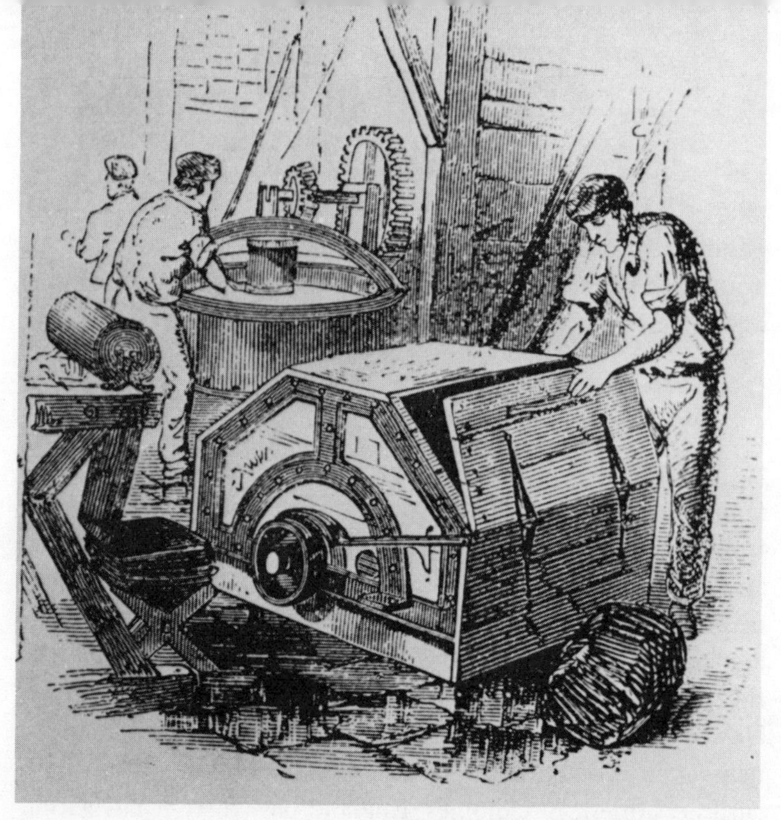

SMITHSONIAN INSTITUTION PRESS

Water is being extracted in this bin in an old-time dyeing plant; at left of the bin, cloth is being rolled up from its folded state.

of the few to compete with vegetable dyes. The process was patented in England in 1840. It called for dipping cloth first in a lead solution and then in a chrome solution. This produced yellow lead chromate.

BROWNS

The heartwood of an acacia tree that grew in Asia and the East Indies contained a gum called cutch that turned cloth a rich brown. It was particularly effective in wool. European and American dyers, however, did not start importing cutch until about 1800. Before then tannin from the bark of hemlocks, maples, and alders produced most of the browns. In colonial America the bark, husks, and nuts of the black walnut and butternut trees produced popular browns. With iron these dyes turned cloth black. Often, using an iron pot was sufficient to turn the color from brown to black.

BLACKS

By varying the dyeing process, logwood, common in many parts of tropical America, gave various colors ranging from blue to black. With a mixture of alum, cream of tartar, and potassium logwood turned wool bluish-gray. Gray could be produced by eliminating the alum. With a simple mordant logwood produced black. In the early struggle for control of the New World the Spanish held a monopoly in logwood. However, England became the rulers in Honduras, found a new supply, and broke Spanish grip on the trade. In the dispute with her American Colonies, England passed strict navigation acts in a vain attempt to keep American ships from carrying logwood and other cargoes to and from non-British ports. The logwood often had additional importance to American merchant captains as ballast for their ships. They ignored the navigation acts, particularly when the difference between a safe and unsafe voyage was at stake.

THE COMING OF SYNTHETICS

In 1828 Germany's brilliant Friedrich Wöhler startled the scientific world by making artificial urea, an organic compound strong in nitrogen that had been found before only in urine. Uuder the concepts of the day, Wöhler's feat was impossible. Chemists believed that the compounds of nature could be made only by nature; they were too complicated to be synthesized in the laboratory. It was foolish to think that man could make such things as sugar, resin, rubber, and dye.

By making urea, Wöhler launched a new science—organic

chemistry. This science is the foundation for the fabulous synthetics industry which today continues to turn out new plastics, paints, dyes, and other products undreamed of in Wöhler's day. Surprisingly, Wöhler turned his back on the science he had founded. He called organic chemistry "a monstrous and boundless thicket." Fortunately, other men dared to explore the thicket. One of them was an 18-year-old college freshman.

William Henry Perkin, known today as the founder of the synthetic dye industry, was extremely fortunate to begin his studies under another brilliant German, August Wilhelm von Hofmann. Hofmann had been invited to England to head the Royal College of Chemistry, a post he held when young Perkin decided to become a chemist. Hofmann gave his students challenging assignments. One of them was to explore the possibility of synthesizing quinine, an antimalarial drug desperately needed in the tropical British colonies. At the time the only source of quinine was the rare cinchona tree. A cheap, synthetic substitute would save thousands of lives.

Perkin decided to work on this assignment over the Easter holiday of 1856. In his home laboratory he began experimenting with toluidine, a derivative of carbon-rich coal tar. He tried mixing the substance with many other chemicals, seeking a reaction that would produce quinine. His experiments failed. He tried working with aniline, another coal tar derivative, but again he failed. All he could produce was a black gum that settled to the bottom of his beakers. A professional chemist probably would have discarded this residue, but Perkin was curious to know what the tar could be. He dried it, ground it to a powder, and dumped it into alcohol. A solution of breath-taking violet appeared before his eyes.

Perkin had studied the Phoenicians and their famous dye. Had he made a dye? He dipped strips of silk into the solution. The silk soaked up the color. When he washed the silk and hung it in the sun, the color remained. He had indeed made a dye.

The rest of Perkin's story is a chronicle of success. Much to Hofmann's disappointment, the young man abandoned his studies. With the help of a brother and his father, a wealthy builder, Perkin built a small factory at Greenford Green, a suburb of London. To extract aniline from coal tar he had to mix benzene, sulfuric acid, and sodium nitrate—the ingredients of dynamite—in big vats. Explosions shook the foundations of the small factory. Fortunately, before anyone was seriously hurt, Perkin learned to control the reactions and began producing his new dye. England at first ignored it, but French textile makers accepted it with enthusiasm, calling the violet "mauve" because of its resemblance to a local flower. Mauve spread quickly throughout Europe. Soon everyone wanted the new color. It launched an era in fashion known as the "Mauve Decade."

Perkin's discovery and spectacular financial success spurred chemists everywhere into experiments with coal tar. New dyes began to appear almost daily. Some produced colors never seen in cloth before. Others were nothing more than the old colors of natural dyes in pure form. But because they were pure and cheap, they replaced the natural dyes almost overnight. There were some exceptions. Sometimes the laboratory processes could not be converted easily to factory-volume production. Indigotin, for instance, the compound that gave indigo, woad, and royal purple its color, was produced in the laboratory early in the 1870s, but it took ten years for practical commercial production to begin.

World War I with its restrictions on trade increased research into synthetics. Germany was hard pressed for fuel, wood, rubber, cloth, and the other necessities of modern life. Her scientists became leaders in discovering new products—

plastics, synthetic fibers, and dyes. After the war English and American scientists often found themselves trailing the German lead. They had to work hard to catch up.

For a time industrialists turned out plastics and other synthetics without full knowledge of their composition and properties. Many of these products failed. Plastic bottles, for instance, might melt in hot water, crack in cold, or dissolve in strong soap. Plastics got a bad name, but today with more advanced knowledge, tough plastics and fibers can be produced to meet harsh demands. Dyes of any color, fadeproof, impervious to the strongest soaps, and often designed especially for specific cloths are taken for granted by modern consumers. New dyes continue to appear almost daily. In fact the synthetic chemists say they have really just begun their exploration of Wöhler's "monstrous and boundless thicket."

MODERN CLASSES OF DYES

Synthetic dyes number literally in the millions. The dyemakers describe them according to chemical composition. Dye-users prefer to describe them according to method of application. The resulting confusion is gradually being cleared with chemical composition dominating as the key to classification.

Azo dyes are by far the biggest chemical class. Azo describes a link in the dye's molecule that is typified by two atoms of nitrogen. There can be one, two, or more of these links in one molecule. Thus we have monoazo, disazo, and polyazo dyes. There are more than two million azo dyes in use today, coloring everything from inks to cloth, and from paper to leather.

Two other large groups are the triphenylmethane and xanthene dyes. These are characterized by complex benzene rings in their linkage. Benzene itself is one of the simplest

U. S. DEPARTMENT OF AGRICULTURE

There are many uses for dyes. These male codling moths, dyed red for identification, have been sexually sterilized by exposure to radioactive cobalt. When released over orchards they mate, but the resulting eggs do not hatch. This is one of many biological types of insect control that are beginning to replace chemical control.

organic compounds, having six atoms of carbon and six of hydrogen joined in an unbroken circle or ring. Xanthene dyes have four benzene rings linked together with atoms of carbon and oxygen. Triphenylmethane dyes have six benzene rings.

Anthraquinone dyes are typified by bonds of carbon and oxygen atoms. Dye-users have subdivided these into acid, mordant, disperse, and vat dyes. Alizarin, a vat anthraquinone dye, is the coloring agent of ancient madder. By using it with an aluminum compound as mordant, dyers can match the old Turkey red.

Sulfur dyes, not yet fully understood by chemists, include a group that dissolves in solutions of sodium sulfide. The dye becomes fixed when it oxidizes in the cloth.

Textile-reactive dyes, among the newest to be developed, create a chemical bond with the fibers they color. Procion dyes, in the textile-reactive class, are available today for home craftsmen.

MAKING DYE

With so many different kinds of dye—we have listed only the major classes—it is impossible to call any factory process typical. No two processes are exactly alike. All involve several steps and all require carefully controlled chemical reactions. In nearly all cases the productifon process begins with a derivative of coal tar or petroleum that manufacturers call the dye intermediary. This is a compound rich in carbon and many of the other necessary elements needed in the end product.

To give some idea of the many complexities involved in modern dye manufacture, let's trace the processes needed in making a popular disazo dye known as Direct Blue 6. The intermediary in this case is benzidine, a substance containing the benzene molecule. It is first poured into a tank lined with

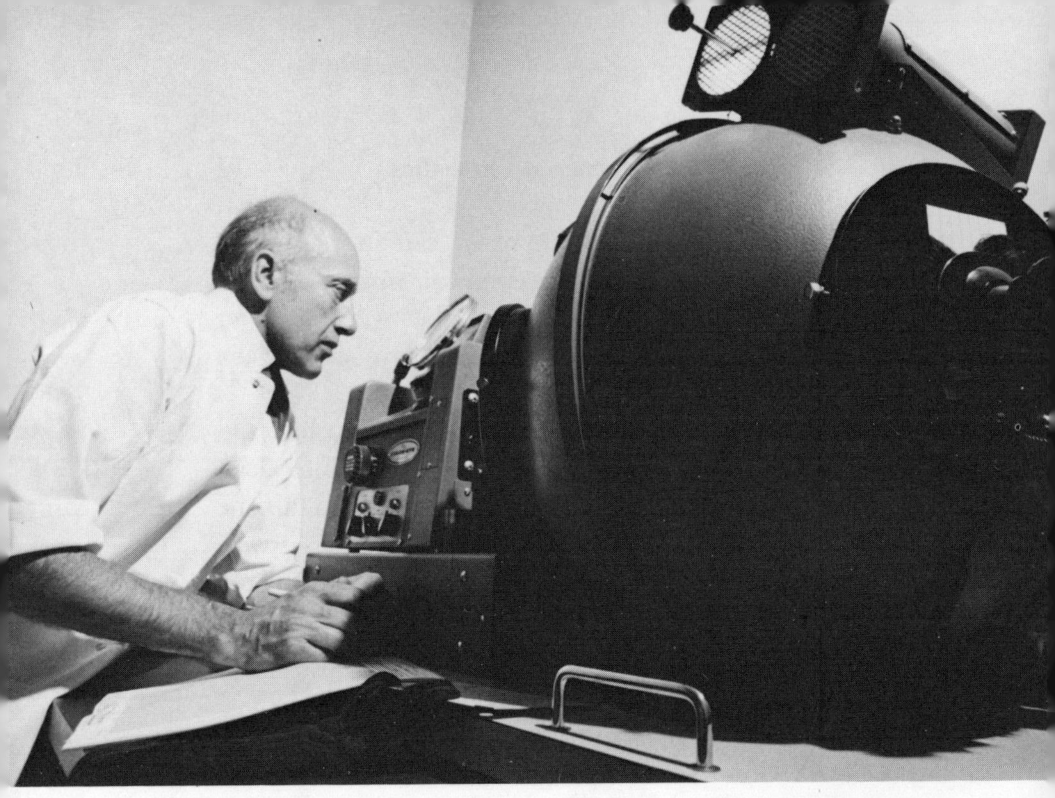

ALLIED CHEMICAL CORP.

Dye manufacturers today control colors to the finest shade. Here a technician uses a machine called the color eye to check a new dye.

wood or brick. The tank must be lined because bare metal would react with benzidine. Motor-driven paddles stir the solution while water and then hydrochloric acid are added. Next, steam jets are opened into the tank, bringing the mixture to a boil. This causes the benzidine to dissolve. When the steam is turned off, workers dump ice into the tank.

As the solution cools, benzidine hydrochloride forms. Gradually, over a period of about two hours, another chemical, sodium nitrate, is run into the mixture. Meanwhile workers prepare a second tank with a mixture of soda ash and acid. The solution from the first tank is then poured into the second. Stirring continues and more ice is added to prevent unwanted reactions that would occur at room temperatures. Gradually, a complex compound containing carbon, hydrogen, sodium, sulfur, oxygen, and nitrogen forms. This is the dye with the characteristic double nitrogen or azo bond. Extracting the compound from the solution requires yet another series of steps. Steam is used once again to heat the mixture. Then common salt is added and the solution is run through a pipe to a filter press. The press squeezes out the unwanted liquid, leaving the dye caked against the cloth folds of the filter. Some filters are equipped with air jets to speed drying.

The cake from the filter then is collected and placed in trays. The trays are placed in driers. Sometimes vacuum ovens are used and sometimes jets of hot air serve to dry the dye. Grinding follows drying, and then the concentrated dye is mixed with salt to bring the dye down to the manufacturer's standard of strength.

Heavy tanks and presses are needed for nearly all dye-making. The equipment is expensive. Sometimes tanks must be covered with sealed lids to get the desired temperature. One dye must be heated to 428° F. to bring about the required reactions. There is nothing easy about making modern dyes but for anyone with an eye for color, the efforts are rewarding.

7
Dye Projects

With home dyeing, whether it is simple tie-dyeing or an ambitious effort to classify the dye plants and their colors in your area, the basic requirements are time and patience. Learn to turn this to your advantage. Use the time to tinker and experiment. You can have as much fun as an alchemist, and the results may give you much more personal satisfaction than a heap of gold.

Before taking up individual projects, let's consider some general rules. With all dyeing projects, you will get the best results with soft water. Serious home dyers go to great trouble collecting rainwater or seeking out good spring water for their solutions. Even dyers lucky enough to have soft tap water will use rainwater if it is available. Tap water is often treated with chemicals which will tone down colors as badly as the calcium of hard water.

Many dyes, particularly the synthetics sold for home use, are toxic. Even the "safe" dyes can cause allergic reactions, including runny noses, sneezes, or skin rashes. Always read manufacturers' instructions on safe handling carefully. If there is any doubt about a dye, treat it as poison. Use rubber gloves unless the label specifically says the dye is nontoxic. The native dyes you can collect from local plants are generally safe, but *never*, no matter what the brew, stand long over boiling dye, inhaling the fumes. Lungs do not take kindly to dyes.

Another general rule with dyes is persistence. Do not give up on a dye too soon. Some work beautifully in one cloth, not at all in another. Also, different mordants often give different results.

You will want to read all you can, particularly if you take up natural dyes. There are many excellent books on the subject. We list our choices in the Suggested Reading list. These books list proven native dyes, but there is one vital thing you must remember. The dye plants vary from one region to another. Even those of the same species can give different results due to climate, soil, or a combination of both. This can give you fits if you let it, but try to think of it as a challenge. The work you do with plants in your region might well be pioneer work.

This leads to a final general rule. When you experiment, and we think you must to appreciate natural dyes fully, keep careful record of the steps you take and the recipes you use. Nothing is more frustrating than producing a wonderful color from plants you have collected and processed yourself and then not being able to remember exactly how you did it. The written record assures repetition of your success.

NATURAL DYEING

Pure woolen yarn will give best results for your first venture with natural dyes. You can use small amounts for tests, and dyed yarn lends itself well to many projects such as knitting, crocheting, hooked rugs, and weaving. Be sure to buy wool free of synthetic fibers, and of course you should start with white wool. Silk takes dye almost as well as wool, but it is generally more expensive. If you are interested in embroidery, however, you can buy white silk embroidery thread at no great expense. Thread lends itself well to testing dyes and is ideal when you are unable to collect large quantities of plant materials. Though cotton will take natural dyes, the colors generally come out paler than they do with wool and silk.

Home dyeing requires two basic steps, mordanting and the dyeing itself. With some dyeing it is possible to put the mordant in with the dye bath and do both steps at the same time, but we do not recommend this for beginners. Besides, for testing dyes you should work with material that has received a standard treatment. This will give you fair comparison in your results. The cloth you mordant can be dried and stored, so it does not hurt to prepare a large batch.

Nearly every drugstore sells alum that is perfectly adequate for dyeing. Into four gallons of warm water put three ounces of alum and one ounce cream of tartar. Stir until the chemicals are completely dissolved. Increasing the alum will not improve results. In fact too much alum will make the finished wool sticky. (Potassium aluminum sulfate, a more complex alum compound available at chemical supply houses, can be used in greater proportion without causing stickiness, but the above recipe is adequate for wool.)

Tie your wool in loose skeins with string. Do not tie the strings too tightly; that will cause uneven mordanting. Thread must be unwound from spools and tied in skeins. Wet the yarn or thread thoroughly in warm water before placing it in the mordant bath. Keep the solution hot for one hour. Do not let it boil. The material is now ready for dyeing. It can go directly into a dye bath or it can be hung in a shady spot to dry. Never dry material in direct sunlight. Avoid sudden changes in temperature, particularly with wool; it will shrink.

Alum, of course, is just one of many mordants. Eventually you will want to experiment with others. In the beginning we recommend you mordant and dye with a porcelain-lined pot, but you can get interesting results with an iron pot, particularly if you want darker colors. A little of the iron from the pot goes into solution and acts as a mordant. Rusty nails in the bottom of the pot will increase the iron content. Goldenrod and many of the other yellow dyes will produce an olive green when used with rusty nails.

Vinegar, ammonia, lye, and tea act as mordants. (Be *very* careful with caustics, such as lye and ammonia.) They are easy to obtain and lend themselves well to testing with small samples at the end of the dye bath. Dip out some of the dye solution into a glass or pan with a short bundle of dyed yarn. Add a few drops of ammonia or vinegar. Often the color will brighten dramatically. Dyers call this "blooming." If the results please you, you can treat the entire dye bath with this supplemental mordant, adding a little at a time until the bloom comes.

A tin mordant (stannous chloride) and a chrome mordant (potassium or sodium dichromate) can be obtained from a chemical supply house or from a craft shop specializing in dyes and textiles. Both are excellent, particularly in brightening yellows and reds. Chrome, however, deteriorates in light. You must keep your pots covered when using it.

DYESTUFF

The general rule of proportion with dyestuff (the original natural substance which, when treated, yields a dye) is one to one: you will need one ounce of dyestuff for every ounce of yarn. Like many rules, however, this one must be broken. You must use two ounces of onion skins, for instance, for each ounce of yarn, and it is not unusual with some leaves and flowers to increase the proportion to four to one. Tough fibers such as bark and roots must be boiled up to two hours or more. Delicate things such as flowers and berries should be simmered or boiled lightly for just a few minutes.

Let's try dyeing some yarn. The dyestuff should be rinsed to remove dirt and other foreign matter. Next place it in a

Paper towels are good for testing patterns in a dip-dye project. You can make handsome scarfs and tablecloths with dip-dyeing.

porcelain, glass, or stainless steel container and cover with water—rainwater or soft water if possible. Allow the dyestuff at least two hours to soak; most home dyers schedule their projects to allow overnight soaking. Then boil or simmer it. If one to two hours of boiling is needed, you must add more water from time to time to replace evaporation loss.

Now strain off the liquid into another pot and allow it to cool. Add the wetted yarn. Bring to a simmer. Most natural dyes need no more than 40 minutes of simmering. Cool and let the yarn soak overnight. If you want to experiment with blooming, or brightening with a mordant, make your tests before removing the yarn. After dyeing, the yarn must be rinsed in cold water. Squeeze it gently to remove excess water until the water runs clear. Never twist it to wring out water. Hang the yarn in the shade to dry.

Before we discuss individual dyestuffs and suggest specific formulas, here is a bit of philosophy about natural dyeing. Remember, you are experimenting. You can learn as much from your failures as from your successes. And yarn that has not taken color can be used again.

FLOWERS

Any flower that will stain your fingers when you crush a petal will work as a dye. Chop up the flowers, put them in a pan, cover with water, and boil or simmer. Light boiling should continue no more than 20 minutes. You can simmer flowers

Marigolds make a rich dye. The flowers should be cut in full bloom, but can be dried and stored in a dry place for a year without losing coloring strength. (During storage they should be checked now and then for mold; discard any moldy flowers and sun the remaining blooms a few hours before storing them again.) The daisies at the left can also be used for dyeing, but don't try to store them.

from one to two hours. Strain the liquid into the dye pot and add enough cold water to cover the wetted fabric or yarn. The dye pot should boil no more than 30 minutes, but you can let it simmer up to two hours.

Goldenrod gives lemon yellow. A gallon container filled with flowers will color one pound of wool. Make bloom tests with chrome, vinegar, or tea.

Marigolds give various depths of golden yellow, according to the color of the flower. The flowers can be picked, dried, and stored for future projects or used fresh. If you boil the flowers, give them no more than 20 minutes on the stove. It is best to simmer the dye bath, but if you choose to let it boil, that too should not continue more than 20 minutes. Alum yarn will come out golden tan. Chrome yarn will be brassy.

Zinnias give light yellow. Boil 15 minutes or simmer an hour. The dye bath should boil no more than 30 minutes. Chrome mordant gives more lasting color than alum when dyeing with zinnias.

Dahlias give yellow with alum and orange with chrome. No boiling should continue more than 20 minutes. Zinnias and dahlias both have flowers of many different colors, but all yield yellow dyes, with the brightest yellows coming from the yellow flowers.

For southern readers, cotton is a possibility. The flowers of the cotton plant give a yellow tan with alum, and brass with chrome. Crush the dried flowers and simmer for an hour. The dye bath can boil 30 minutes. When dyeing with cotton, it is often necessary to bloom the bath with vinegar or chrome to get best results.

LEAVES

Most dyers recommend soaking leaves in water overnight before bringing them to a boil. Boiling takes from twenty minutes to an hour for soaked leaves. You can judge the boiling time by observing the color of the liquid. Alder, white birch, mountain laurel, and poplar leaves give yellow. Alder and birch work best with alum mordant. Laurel and poplar work best with chrome. Sage leaves and twigs turn yarn greenish yellow, but you must use a proportion of three to one. It takes three pounds of leaves and twigs to color one pound of wool. Tea leaves turn yarn rose-brown. You will need eight ounces of black tea for one pound of wool, and you do not have to soak tea leaves overnight. Simply boil them for 15 minutes and strain the liquid into the dye pot. Rusty nails in the dye pot will darken the color.

BARK

Bark chips must have long soaking, overnight at least, and they must be boiled about two hours to extract their colors. Careful straining is necessary to keep splinters out of the dye bath. Generally you should use the inner bark. The older, outer bark has little color. Apple, birch, hickory, and oak barks give yellowish tans. A richer brown can be achieved with rusty nails in the dye bath. The dye bath from barks should be be boiled 30 minutes. Barks gathered in the fall or winter have the strongest color.

Barks present a problem because of their high tannin content. The tannin will darken the cloth with age. To prevent this make up a solution of four gallons of water, seven tablespoons of vinegar, and a sixth ounce of potassium chromate. Transfer the yarn from the dye bath into this solution and boil for ten minutes. This will remove much of the tannin from the cloth. If you use an iron pot for dyeing or have placed rusty nails in the dye bath to get darker colors, you will not need the vinegar and chrome solution. Iron also removes tannin.

BERRIES AND NUTS

Blackberries will give pink; privet berries will give blue-green; juniper berries will give khaki, and concord grapes will give purple. The time needed to extract the dye depends on their ripeness and the natural consistency of the berries. Ripe blackberries, for instance, yield their color with gentle crushing and brief simmering. Juniper berries, on the other hand, must be crushed, soaked overnight, and boiled at least an hour. Boiling of the dye bath itself takes from one to two hours to impart color to the cloth. The rule on flowers holds true for berries. If they will stain your hands, they will yield a dye.

Walnut hulls provided American colonists with their most practical brown dye. Without mordant the hulls give a light brown with fair resistance to fading. With alum mordant they give a permanent dark brown. The black walnut hulls give darker browns than the English walnut. You can use the green hulls available in late summer or you can collect them dry in the autumn. The hulls store well without loss of dyeing power. Figure on a pound of hulls for every pound of wool. An hour of soaking and another hour of boiling will extract the color. Hickory nut hulls cut up, soaked overnight, and boiled for 45 minutes will make a dye bath that will turn wool with alum light brown. Pecan hulls, boiled for 15 minutes, will make a dye bath turning wool with alum brown.

ROOTS

Roots of the hollygrape, or Oregon grape, and of the barberry yield a dye called berberine. It will give a buff color to wool with alum. The roots must be chopped, soaked, and boiled before the color can be extracted. About two hours of boiling is necessary, and another half hour will be needed to impart color to the cloth. Sassafras, common throughout the eastern states, has a root that gives a brown to wool with alum and a rose-brown to wool used with chrome. The chopped roots must be soaked overnight and boiled for 30 minutes. Another 30 minutes of boiling is needed for the dye bath.

LICHENS

Lichens for dyes can become a full-time study and occupation. Simply learning to identify the lichens in your area usually becomes a challenge. This unusual plant is actually two plants, an alga and a fungus living together in what scientists call a symbiotic state. They grow on rocks and tree trunks and have flat leathery leaves of brown or greenish gray. The Spanish moss that hangs from cypress and oak trees is actually a lichen.

Dyers separate lichens into two kinds, those that yield their dye through boiling and those that yield it through fermentation in strong solutions of alkali. This second group gave the orchil dyes of olden times. You can get browns, yellows, reds, and purples from lichens. To find out what color a lichen will give, soak a few of the leaves in a strong solution of ammonia and water for a few days. The lichens that yield their dye through boiling can be used in much the same way you would use tree leaves. Orchil lichens must be fermented from three weeks to a month in a solution of ammonia and water. This is a lot of trouble to take for a dye, but the resulting red or purple color can be extremely rewarding. If you are interested in using lichens, we urge you to read Eileen M. Bolton's book *Lichens for Vegetable Dyeing,* which is listed in the Suggested Reading list.

This young craftsman wants to find out if the lichen on this oak is the type that will make a purple dye. A few leaves of it are put in a strong ammonia solution for a few days. It works. Now he can collect a large quantity of lichen and let it ferment in ammonia solution for two weeks. He will have a fine dye once much used by the ancients.

Onion skins (which can perhaps be put aside for you by your local grocer) make a good dye for a winter project when you cannot collect growing natural dyestuff.

KITCHEN DYES

We have mentioned tea leaves, but most of the other dyestuffs described so far require excursions into the countryside or at least the garden. During winter days when you cannot get out, try dyeing with onion skins, coffee, or turmeric.

You can make a golden yellow dye bath by soaking and boiling ordinary onions skins. Red onions will give a rich orange. Use only the outer, dry skin. For every weight of wool, you will need twice the weight of skins, so it might be a good idea to ask that skins be saved for you at the local market.

Turmeric, a common condiment in most kitchens, provided an early yellow dye. The color is good, particularly in wool, but it will fade in strong and prolonged sunlight.

Instead of throwing out the coffee grounds each day, save

them for dyeing. You should use about two pounds of dry grounds for every pound of wool. Boil the grounds for 20 minutes, strain them out, and mix the liquid with water for your dyebath. The dyebath should boil about 30 minutes. Alum will give light brown. Chrome will bring out more yellow in the wool.

OTHER HOME DYES

Certainly you will get most satisfaction from making dyes out of dyestuffs you have collected yourself, but there are other time-saving dyes that work well in home projects. The all-purpose textile dyes are probably the easiest to use. Rit is probably the most common brand name. These are available as either liquids or powders. Both give good results when permanence is not required. They lend themselves well to tie-dyeing of cheap cottons. The all-purpose dyes, however, will fade with washing.

The dye industry, aware of the increasing number of home craftsmen, has begun to produce long-lasting dyes that can be used in knitting, weaving, and other fabric projects. There are available acid dyes and fiber-reactive dyes, the Procion dyes. The latter, developed in England, cannot be matched for permanent results. These new dyes can usually be purchased through craft shops specializing in textiles. The dyes have some drawbacks. For one thing they are expensive, far more expensive than dyes you would make yourself. Also, some of them are poisonous. You must always follow the directions for use and cautions for safety.

More and more craft shops are stocking or have access to the traditional natural dyes such as logwood, fustic, and cochineal. These too are expensive, but if you want to perfect your understanding of the use of these dyes and the appearance of their colors, here is by far the best opportunity.

TRICKS AND PATTERNS

Home dyeing, of course, lends itself to any craft using or producing textiles. Embroidery thread, knitting and weaving yarn, prewoven cloth, even paper can be dyed. Craftsmen working with wool today often dye it as fleece. Then they card and spin the wool for their project. This is a return to old ways that marked quality materials. It is painstaking but satisfactory work. Almost always any work you do with dye will give you pride in producing something no one else can match, something that is yours alone. Here are just a few of the things you can try.

DIP-DYEING

Anyone who has made paper cutouts will have no trouble finding success with dip-dyeing. The same principles are involved. With cutouts you use scissors to make your pattern after folding the paper. With dip-dyeing you use color to make your pattern. This can be done with paper or cloth. In fact it is a good idea to test various folds with paper toweling before using more expensive cloth or paper.

You can use several colors in a dip-dyeing project. Dip one corner of the folded white cloth in red and another corner in blue. You will have a pattern that is both wild and patriotic. If you plan to change the fold between colors, you must let the cloth or paper dry before refolding it. Do not expect sharp edges between the dyed and undyed portions of your material. Most textiles soak up dye from the bath, causing a fuzzy edge. This can be reduced somewhat by wetting the material thoroughly in cold water before you fold and dip it. Incidentally, rice paper, though expensive, works very well in dye. You may produce some patterns worth framing.

Tie-dyeing with rubber bands is simple. The tighter you get the rubber, the whiter the cloth beneath it will appear after dyeing. Cotton T-shirts are ideal; don't start with more expensive material. An enamel-coated pot is excellent for most dye projects. Dip-dyeing and tie-dyeing are easily combined.

A T-shirt gets a cosmic feeling from the tie-dyeing. The tightest rubber band made the brightest ring, in center; the pale outer ring resulted from the loosest tie.

TIE-DYEING

String, cord, ribbons, and tough rubber bands are the key tools for tie-dyeing. By binding the cloth tightly in certain areas, you prevent or retard dyeing in these areas and thus produce a pattern. As with all dyeing, the material must be cleaned and premoistened before dyeing. If you are working with a ready-made garment, backing and labels must first be removed. Though the principle of tie-dyeing is simple, careful planning will give most satisfaction. The most pleasing results come from sharp contrast in the pattern between the dyed and undyed areas of the cloth. Many dyers try to use too many colors or patterns that are too intricate. The contrasts are lost.

You can produce marbled patterns by random crumpling of the cloth. When the cloth is crumpled into a ball, tie it well with your string or rubber bands. Spirals can be produced by twisting the cloth tightly before you tie it. Parallel lines come from even pleating. You can also produce many different random patterns by knotting the cloth. Stars and sunbursts can be made by bunching and tying. Try binding various objects such as shells, buttons, pieces of wood, and corks into the cloth.

Interesting results come through combining dip and tie-dyeing techniques. Don't be afraid to experiment.

BATIK

With tie-dying, even with careful planning the results cannot be predicted. This is part of the fun, but if you want more control of your patterns, you will want to try batik. Intricate, realistic pictures can be dyed in cloth with batik, particularly on silk, but the technique also lends itself to casual patterns for clothing and wall hangings.

Here hot wax is the key tool. You should use a frame to stretch your cloth as you prepare it, and you should put down a generous layer of newspapers under the frame. Paraffin can be melted in a double boiler or in a can set in a pan of hot water. Use a brush to apply the hot wax to the cloth. Make sure that the wax goes right through the cloth. It is a good idea to sketch your design on the cloth first with colored chalk and then cover the chalked areas with wax.

The wax, of course, serves as a dye shield. The areas you cover with it will not take color. Obviously you have to use cold dye baths. Boiling would melt the wax and destroy your pattern. Crumpling the cloth lightly before you put it in the dye bath cracks the wax, and this results in thin, charming lines of color, a unique trait of batik.

Patterns of several colors can be made with this technique. You can either remove all the wax and brush on a new pattern between each dye bath or you can leave the old wax and add to it with your brush before putting the cloth into the new dye bath. Wax can be removed either by ironing the cloth between sheets of newspaper or by dipping the cloth into boiling water.

Wax can be removed from brushes fairly well with turpentine and soap, but we suggest you use old or cheap brushes for this project. Hot wax is not kind to brush bristles.

THE BEGINNING

Limited space has forced us to treat dyeing as an isolated study or craft. Actually, it is almost always part of a larger undertaking. Knowing what colors you can extract from plants, for instance, is just part of the broad study of botany. For crocheters, knitters, and weavers, dyeing is just the beginning. You can make a hooked rug or a tapestry with your yarn. You can weave with yarn that has been tie-dyed or dip-dyed or with yarn you have colored with dyes from plants around your home. Incidentally, weavers through the centuries have produced unique patterns with tie-dyed yarns. The colors, of course, have come from available plants or animals.

In short, the art of dyeing leads to other things, many other things. While mastering the art brings satisfaction in itself, do not neglect it as the beginning for other arts and other skills.

Glossary

absorbed light That portion of light which is neither reflected nor filtered by an object.

additive mixing A term used to describe the combination of one or more filtered lights.

alum A compound combining sulfur, aluminum, potassium and other elements, long used as a mordant in dyeing.

amethyst Violet or reddish-violet quartz, a precious stone worn by Egyptian warriors in the belief that it gave courage.

aquatint An acid engraving process used to produce half-tones in etchings.

atom The smallest chemical division of matter.

casein A protein compound in milk with strong binding property, an important binder for paints.

chromogen The chemical portion of a pigment or dye which determines its color.

compound A chemical which chemically combines two or more atoms of different elements.

cones Those cells in the retina of the eye which distinguish color.

copperas Ferrous sulfate, a compound used extensively in making writing inks.

direct light That light which comes from an energy source such as the sun, the glowing filament of a light bulb, or a burning candle.

dry point An etching technique in which a sharp instrument is used to scratch lines in a printing plate.

dye A coloring agent which dissolves in solution and will impart the color to any substance which will absorb the solution.

dye intermediary An industrial term used to describe compounds containing the basic ingredients of dyes.

dyestuff Original natural substance which, when treated, yields a dye.

electromagnetic effect The phenomenon exhibited by some metals which produces electric currents when exposed to light.

element A chemical substance having no more than one kind of atom. There are more than 100 known elements.

enamel A glossy paint using varnish as the vehicle.

encaustic Hot-wax painting.

extender pigment A pigment selected more for its consistency than its color. Extenders are sometimes called inert pigments.

filtered light That light which passes through an object.

flat-bed press A printing press in which the plate is laid on a flat surface and the matter to be printed is rolled over or pressed against it.

fresco Painting on wet plaster.

gallotannate ink An ink utilizing the property of some iron compounds to turn brownish-black when combined with weak acids.

gravure A printing method using plates that carry ink in depressions in the plate surface.

gum arabic A paint binder obtained from secretions of a species of Asiatic acacia.

hectograph A printing method using an ink-absorbing gelatin.

heliotype A printing method using a photosensitive gelatin applied to a glass plate.

indirect light Light reflected from objects. Moonlight is a classic example of indirect light.

jaundice A disease of the liver which causes yellowing of the eyes and skin.

lacquer Dissolved nitrocellulose. Natural lacquer from sumac resin was first used by the Chinese.

lake A pigment colored by a dye.

letterpress A printing method in which the ink is transferred to the paper from raised portions of the printing plate. Also called relief prniting.

lithography A printing method using a flat plate, portions of which have been treated to accept ink and other portions to reject it. Limestone was used originally, but commercial lithography is now done with metal plates.

mimeograph A form of stencil printing widely used in offices; also, any duplicating machine of this type. Wax stencils for the mimeograph can be cut easily on an ordinary typewriter.

molecule Two or more atoms bonded together; the smallest unit of a compound.

monoprint A print taken from a hand-inked plate, usually glass. No second copy is possible.

mordant A chemical, usually a metallic salt, that bonds a dye to fibers. In some combinations of dye and fibers, mordants also influence hue and color intensity.

offset A printing method in which ink is transferred to a blank roller which in turn carries the ink to paper.

opaque A term describing objects which let no light pass through them.

organic Complex carbon atoms are organic compounds. The term originally referred only to such compounds produced by nature, but it now refers to both natural and man-made carbon compounds.

oxidation A chemical reaction involving the addition of the oxygen atom.

pigment The solid particles which give body and color to paints and inks.

photogravure A gravure method in which the plate is prepared by photography and chemical etching.

photon A minute unit (or "particle") of energy by which light travels, as well as by waves. Though sometimes called a "particle," it is not a unit of matter.

polymer A molecule of large size formed of a repeating pattern of many relatively simple molecules. Polymers can change their structure without changing their chemical composition.

prism A triangular piece of glass which bends light, separating the various wavelengths to produce the visible spectrum.

reducing agent A chemical which makes an insoluble compound soluble.

reflected light That light which bounces off an object.

relief printing See *letterpress.*

retina The back inner surface of the eye, made up of light-sensitive cells.

resin The sticky secretion of plants and trees which provides the base for varnishes and enamels. It can now be produced synthetically.

rods Those cells in the retina of the eye which distinguish white light and dark.

rosin The hardened resin of certain pine trees.

rotary press A printing press in which the plates are carried on cylinders or drums.

rotogravure A rotary press using gravure plates; also the printed matter from this press.

shaman A medicine man; a priest-doctor using magic.

shellac A coating made from dissolving a resin secreted by the lac insect.

stain A wood coating using penetrating oils as the vehicle and dyes or pigments as coloring agents.

stencil printing A printing method using a thin surface in which portions have been cut away to allow passage of ink onto paper, cloth, or some other substance.

subtractive mixing Combining of two or more light-absorbing coloring agents such as pigments.

tempera A paint using a water-soluble binder such as egg, glue, or casein.

translucent A term for an object which allows the passage of some but not all light.

transparent A term for an object which allows the passage of all or nearly all light.

urea A compound found in animal urine. The first organic compound to be synthesized.

valence The bonding capacity of atoms, or the tendency of element atoms to form compounds. Schematically, valence can be thought of as attracting arms surrounding an atom. A hydrogen atom, for instance, has one arm (valence: 1) an oxygen atom two, and a carbon atom four.

varnish A dissolved resin used either as a clear coating or as the vehicle for pigmented paint.

vehicle The liquid portion of paints and inks.

visible spectrum That portion of electromagnetic energy which the eye can see.

Suggested Reading

BOOKS

Adler, Irving, *Color in Your Life* (New York, John Day Company, 1962)

Adrosko, Rita J., *Natural Dyes of the United States* (Washington, D.C., Smithsonian Institution Press, 1968)

———, *Natural Dyes and Home Dyeing* (New York, Dover Publications, 1973)

Allen, Agnes, *The Story of the Book* (London, Faber and Faber, no date)

Arnold, Grant, *Creative Lithography and How to Do It* (New York, Harper and Brothers, 1941)

Asimov, Isaac, *The Search for the Elements* (New York, Basic Books, 1962)

Baldwin, Gordon C., *Talking Drums to Written Word* (New York, W. W. Norton and Co., 1970)

Barnes, Albert C., *The Art in Painting* (New York, Harcourt, Brace and Co., 1937)

Bauman, Hans, *The Caves of the Great Hunters*, trans. by Isabel and Florence McHugh (New York, Pantheon Books, 1954)

Beeler, Nelson F. and Branley, Franklyn M., *Experiments with Light* (New York, Thomas Y. Crowell Company, 1964)

Bemiss, Elijah, *The Dyer's Companion* (New York, Dover Publications, 1973)

Birren, Faber, *Color, a Survey in Words and Pictures* (New Hyde Park, N.Y., University Books, 1963)

Bolton, Eileen M. *Lichens for Vegetable Dyeing* (London, Studio Books, 1960)

Boy Scouts of America, *Painting* (New Brunswick, N.J., Boy Scouts of America, 1954)

Brooklyn Botanic Garden Editorial Committee, *Dye Plants and Dyeing—A Handbook* (Brooklyn, N.Y., Brooklyn Botanic Garden, 1964)

Bryan, Nonabah G. and Young, Stella, *Navajo Native Dyes* (Lawrence, Kans., U. S. Department of Interior, Bureau of Indian Affairs, 1940)

Burman, Harold G., *Principles of General Chemistry* (Boston, Allyn and Bacon, 1968)

Cobb, Hubbard H., *How to Paint Anything* (New York, The Macmillan Co., 1972)

Cooper, Elizabeth K., *Discovering Chemistry* (New York, Harcourt, Brace and Co., 1959)

Doerner, Max, *The Materials of the Artist*, rev. ed., trans. by Eugen Neuhaus (New York, Harcourt, Brace and Co., 1949)

Fisk, Phillip M., *The Physical Chemistry of Paints* (New York, Chemical Publishing Co., 1965)

Fraser, Douglas, *Primitive Art* (Garden City, N.Y., Doubleday and Co., 1962)

Freeman, Ira M., *Light and Radiation* (New York, Random House, 1970)

Freeman, Mae and Ira, *Fun and Experiments with Light* (New York, Random House, 1963)

Gardner, Percy, *A Grammar of Greek Art* (New York, The Macmillan Co., 1905.

Giedion, S., *The Eternal Present: The Beginnings of Art* (New York, Bollingen Foundation, 1962)

Hafner, German, *Art of Rome, Etruria and Magna Graecia* (New York, Harry N. Abrams, 1969)

Harrison, George Russell, *The First Book of Light* (New York, Franklin Watts, 1962)

Hellman, Hal, *The Art and Science of Color* (New York, McGraw-Hill Book Co., 1967)

Hicks, Ami Mali and Oglesby, Catharine, *Color in Action* (New York, Funk & Wagnalls, 1937)

Higgins, Reynold, *Minoan and Mycenaean Art* (New York, Frederick A. Praeger, 1967)

Hiler, Hilaire, *The Painter's Pocket-Book of Methods and Materials* (New York, Harcourt, Brace and Co., no date)

Homann-Wedeking, E., *The Art of Archaic Greece* trans. by J. R. Foster (New York, Crown Publishers, 1966)

Howell, F. Clark, *Early Man* (New York, Time-Life Books, 1968)

Hyde, Margaret O. and Hyde, Bruce G., *Atoms Today and Tomorrow* (New York, McGraw-Hill Book Co., 1970)

Janson, H. W. with Cauman, Samuel, *History of Art for Young People* (New York, Harry N. Abrams, 1971)

Jensen, Hans, *Sign, Symbol and Script*, trans. by George Unwin (New York, G. P. Putnam's Sons, 1969)

Judd, Deane B. and Wyszecki, Günter, *Color in Business, Science and Industry* (New York, John Wiley and Sons, 1968)

Kent, James A., ed., *Riegel's Industrial Chemistry* (New York, Reinhold Publishing Corp., 1962)

Lesch, Alma, *Vegetable Dyeing* (New York, Watson-Guptill Publications, 1970)

Mark, Herman F. and the editors of Life, *Giant Molecules* (New York, Time Inc., 1966)

Martini, Herbert E., *Color* (Pelham, N. Y., Bridgman Publishers, 1946)

Mayer, Ralph, *The Artist's Handbook of Materials and Techniques*, rev. ed. (New York, Viking Press, 1970)

———, *The Painter's Craft* (New York, D. Van Nostrand Co., 1966)

Morrison, A. Cressy, *Man in a Chemical World* (New York, Charles Scribner's Sons, 1937)

Moulin, Raoul-Jean, *Prehistoric Painting*, trans. by Anthony Rhodes (New York, Funk and Wagnalls, 1965)

Mueller, Conrad G., Rudolph, Mae, and the editors of Time-Life Books, *Light and Vision* (New York, Time-Life Books 1966)

Ogg, Oscar, *The 26 Letters* (New York, Thomas Y. Crowell Co., 1971)

Paschel, Herbert P., *The First Book of Color* (New York, Franklin Watts, 1959)

Powell, T. G. E., *Prehistoric Art* (New York, Frederick A. Praeger, 1966)

Purdy, Susan, *Holiday Cards for You to Make* (Philadelphia, J. B. Lippincott Co., 1967)

Rublowsky, John, *Light, Our Bridge to the Stars* (New York, Basic Books, 1964)

Ruchlis, Hy, *The Wonder of Light, a Picture Story of How and Why We See* (New York, Harper and Brothers, 1960)

Selwood, P. W., *General Chemistry*, fourth ed. (New York, Holt, Rinehart and Winston, 1965)

Sootin, Harry, *Light Experiments* (New York, W. W. Norton & Co., 1963)

Wise, William H., *Home Painting, Wallpapering and Decorating* (New York, William H. Wise & Co., 1955)

Wohlrabe, Raymond A., *Exploring Giant Molecules* (New York, The World Publishing Co., 1969)

Wolfe, Herbert Jay, *Printing and Litho Inks*, sixth ed. (New York, MacNair-Dorland Company, 1967)

Woody, Russell O., Jr., *Painting with Synthetic Media* (New York, Reinhold Publishing Corp., 1965)

ARTICLES

Consumer Bulletin staff, "Writing Inks," *Consumer Bulletin*, October 1969

Daly, Margaret Vennum, "Invisible Ink in Color, a New Way to Help Your Kids Learn," *Better Homes*, May 1968

Hogarth, Paul, "Drawing with Reeds and Quills," *American Artist*, September 1968

Melvin, A. Gordon, "Dye from Shells," *Hobbies*, February 1971

Skelton, Marion A., "A Time to Dye," *Science and Children*, November, 1969

INDEX